Survival: One Health, One Planet, One Future

Planet Earth has been here for over 4.5 billion years but in just two human generations we have managed to place our only 'home' at great risk. Many lessons from history have not yet been learned and new lessons may prove equally, if not more, difficult to take on board as we head deeper into the twenty-first century.

This book highlights two of our greatest social problems: changing the way we relate to the planet and to one another, and confronting how we use technology (dataism) for the benefit of both humankind and the planet. Covering a wide range of key topics, including environmental degradation, modern life, capitalism, robotics, financing of war (vs peace) and the pressing need to re-orient society towards a sustainable future, the book contends that lifelong learning for sustainability is key to our survival. The author argues that *One Health* – recognising the fundamental interconnections between people, animals, plants, the environment – needs to inform the UN-2030 Sustainable Development Goals and that working towards the adoption of a new mindset is essential. We need to replace our current view of limitless resources, exploitation, competition and conflict with one that respects the sanctity of life and strives towards well-being for all, shared prosperity and social stability.

Clearly written, evidence based and transdisciplinary – and including contributions from the World Bank, InterAction Council, Chatham House, UNESCO, World Economic Forum, the Tripartite One Health collaboration (UN Food and Agriculture Organization, World Organisation for Animal Health and World Health Organization), One Health Commission and more – this book cuts across sociopolitical, economic and environmental lines. It will be of great interest to practitioners, academics, policy-makers, students, non-government agencies, corporations and the public at large in both developed and developing nations.

George R. Lueddeke MEd PhD is an educational advisor in higher and medical education and chairs the global One Health Education Task Force for the One Health Commission and the One Health Initiative. He has published widely on educational transformation, innovation and leadership and been invited as a plenary speaker to different corners of the world.

Routledge Studies in Sustainability

Personal Sustainability
Exploring the Far Side of Sustainable Development
Edited by Oliver Parodi and Kaidi Tamm

Sustainable Modernity
The Nordic Model and Beyond
Edited by Nina Witoszek and Atle Midttun

Towards the Ethics of a Green Future
The Theory and Practice of Human Rights for Future People
Edited by Marcus Düwell, Gerhard Bos and Naomi van Steenbergen

Sustainability Transitions in South Africa
Edited by Najma Mohamed

Strongly Sustainable Societies
Organising Human Activities on a Hot and Full Earth
Edited by Karl Johan Bonnedahl and Pasi Heikkurinen

Managing Complexity, Earth Systems and Strategies for the Future
Walter R. Erdelen and Jacques G Richardson

Measuring Intangible Values
Rethinking How to Evaluate Socially Beneficial Actions
Marie Harder and Gemma Burford

Education for Sustainable Human and Environmental Systems
From Theory to Practice
Edited by Will Focht, Michael A. Reiter, Paul A. Barresi and Richard C. Smardon

Survival: One Health, One Planet, One Future
George R. Lueddeke

www.routledge.com/Routledge-Studies-in-Sustainability/book-series/RSSTY

Survival: One Health, One Planet, One Future

George R. Lueddeke

Routledge
Taylor & Francis Group
LONDON AND NEW YORK

earthscan
from Routledge

First published 2019 by Routledge

2 Park Square, Milton Park, Abingdon, Oxfordshire OX14 4RN
52 Vanderbilt Avenue, New York, NY 10017

Routledge is an imprint of the Taylor & Francis Group, an informa business

First issued in paperback 2020

British Library Cataloguing-in-Publication Data
A catalogue record for this book is available from the British Library

Library of Congress Cataloging-in-Publication Data
A catalog record has been requested for this book

ISBN: 978-1-138-33495-3 (hbk)
ISBN: 978-0-367-50063-4 (pbk)

Typeset in Goudy
by Wearset Ltd, Boldon, Tyne and Wear

To all those who believe
in the Oneness of the world

Contents

Figures

Tables

Boxes

Foreword

Like many of my colleagues, I get up each morning and read the news, listen to the commentators, and learn a bit more about the risks associated with life in today's world to us as individuals and to the environment in which we live. I then get on with my daily life, sometimes placing a mental bookmark on the challenges we face, to which I intend to return one day in order to gain better understanding – and I store them in a mental to-do file.

Having read Dr George Lueddeke's well researched, lucid and absorbing book about our survival – a major theme that underpins the UN-2030 Sustainable Development Goals (SDGs) – it has suddenly become more difficult to place those risks in the to-do file and to deal with them perhaps at a later time. The urgency of dealing with them now becomes a clear imperative. As the book begins, the author pulls together the fragments of the various threats to our health and well-being and to our environment that we read and hear about each day. They vary from risks to human and animal health from plastics and the waste they create to the consequences from climate change and our own dietary and eating habits, and how the risks impact on the seemingly insurmountable inequalities within and between our societies. It becomes a book of current events and twenty-first century trends that will impact on our health and well-being, our planet, our future and the future of our children.

The global issues that the author highlights – poverty, inequality, health, modernity, extremism, war and peace, migration, education and climate change, to name but a few – are all interconnected in an interdependent world and must be tackled together. These issues and others are succinctly described in the book, as is the evidence that confirms them. They are presented in full detail, interspersed with quotes and anecdotes from the world's former and present leaders in science and politics. Reading the initial chapters is uncomfortable – they amplify and make more real that which we already know and, ironically, include evidence about new technologies such as artificial intelligence that provide solutions to mitigate some of the risks, but that create new risks in themselves in a world increasingly driven by technology. There is much to grasp – and much to contemplate as we strive to make a healthy and more sustainable planet.

Early in the book it also becomes clear that mitigation of the risks we face is no longer a question of what or why but *when and how*. To this end, the author convincingly argues that it requires understanding and taking personal responsibility for our actions – *supported by educational opportunities from early years to lifelong learning in both developing and economically more advanced nations*. He takes pains to demonstrate how lessening external threats necessitates that action also be adopted collectively no matter where we live, or what we believe. From its almost resigned and gloomy beginning – where risks that are preventable are shown to be present in all aspects of our daily lives – the book gradually moves towards a blueprint of what needs to be done to allay these risks and sustain our planet. Indeed, it allocates an entire chapter to contributions by leading global organisations and major movements and the collective actions underway to support their initiatives – reasons for guarded optimism.

In *Survival*, George Lueddeke argues compellingly for a new worldview that will ensure that our needs as human beings become compatible with the needs required to sustain our ecosystem. His vision reflects the *One Health and Well-Being concept* – a 'global unity of purpose' requiring joint action between human and animal health and the ecosystem – as the key to creating an environment where people from all walks of life work together to ensure the sustainability of the needs for human survival, and of those of the planet.

The last few chapters and insightful Epilogue summarise the author's ten 'propositions for global sustainability' and underscore the key role of future generations that must be motivated 'to make the world work better'.

Survival by George Lueddeke is an important read about what it takes to lessen the dangers to health and well-being, and to ensure a sustainable planet. It challenges us directly to spread the message across the social and economic divide – to those of us who create risks for corporate or personal vested interest, and to those who cannot easily access the knowledge but must do so in order to sustain their future.

One of the book's concluding comments underscores the urgency for transformative change – continued and increased collective action – in particular, recognising the importance of global responsibility for the prevention of nuclear war, global warming and genetically engineered viruses. If we fail, the author cautions, 'the shelf life of *Homo sapiens* could be extremely short'. *Survival* is a timely wake-up call for us all.

David L Heymann MD
Centre on Global Health Security
Chatham House (London)

Preface

Averting apocalypse in the twenty-first century

Yuval Noah Harari, author of *Homo Deus: A Brief History of Tomorrw*[1] offers a glimpse into the future. One of the scenarios presents a rather chilling prospect whereby circumstances might lead us to a dysfunctional or dystopian world characterised by (1) a scientific dogma that says 'life is data processing'; (2) decoupling of intelligence from consciousness (feelings); and (3) the possibility that highly intelligent algorithms (superhuman or artificial?) will 'know us better than we know ourselves' (essentially controlling our lives) – reality vs science fiction?

Alternatively, because 'technology is not deterministic,' Harari also proposes that: 'The same technology could create very different kinds of societies' as we have seen with regard to the socioeconomic and geopolitical outcomes of 'the Industrial Revolution'. To avoid reaching hasty conclusions, he reassures us that 'All the scenarios outlined' in *Homo Deus* 'should be understood as possibilities rather than prophecies' and he encourages us 'to think and behave in new ways that will prevent these particular possibilities from materializing'.

My present book, which complements *Global Population Health and Well-Being in the 21st Century*,[2] aims to respond to Harari's challenge. It confronts the longer-term threat posed by drifting into a technocratic fog devoid of human thought and emotion.

Survival: One Health, One Planet, One Future argues for the existential importance of standing up for the sustainability of the planet and humanity. It advocates the adoption of a new mindset: replacing the current view of limitless resources and our huge appetite for exploitation, competition and conflict with one that respects the sanctity of life and strives towards well-being, shared prosperity and social stability.

The book is evidence-based, and highlights that two of the greatest social problems of this century are changing the way we relate to the planet and to each other and confronting how we use technology for the benefit of both humankind and the planet. The key concept that underpins the book is *One Health*, which essentially refers to 'the need for multiple disciplines or professions to work together – locally, nationally and globally – to solve complex problems and with a view to attaining optimal health and well-being for people, animals and the environment'.[3] Throughout the book, *One Health* is considered to be the unifying thread that may give us the best chance of sustaining humanity in the decades ahead, especially as we are increasingly being pulled into a cyberworld governed by advanced algorithms that may or may not be of our own making.

Ultimately, *Survival: One Health, One Planet, One Future* argues for a world which continues to build and maintain well-functioning and empowered societies and that places collaboration over division, sustainability over short-term gain, humanity over technology and peace over destruction.

Intended for a wide-ranging audience that crosses social, economic, political and environmental sectors and from more economically developed as well as developing nations, the book's main purpose is to clarify, inform and stimulate discussion and debate on global and national issues. Above all, it seeks to offer hope for a brighter future – for people and the planet – especially given the times in which we find ourselves.

George Lueddeke PhD
August 17, 2018

References

1 Harari YN. *Homo Deus: A Brief History of Tomorrow*. London: Harvill Secker (Vintage); 2016.
2 Lueddeke GR. *Global Population Health and Well-Being in the 21st Century: Toward New Paradigms, Policy and Practice*. New York: Springer Publications; 2016.
3 One Health Commission. *What is One Health?* www.onehealthcommission.org/en/why_one_health/what_is_one_health/ (accessed 4 July 2018).

Acknowledgements

Sincere appreciation is extended to members of the *One Health Commission* (OHC) and the *One Health Initiative* (OHI) for allowing me to become part of the *One Health* 'community'. In particular, at the OHC I wish to thank Joann Lindenmayer, Cheryl Stroud, Gretchen Kaufman and Laura Schoenle; and, at the OHI, Bruce Kaplan, Laura Kahn, Thomas Monath and Lisa Conti for helping to progress the *One Health and Well-Being* movement globally.

I am indebted to David Heymann MD, professor of Infectious Disease Epidemiology, London School of Hygiene & Tropical Medicine; head of the Centre on Global Health Security at Chatham House; and chair of Public Health England, for offering to write the Foreword for *Survival: One Health, One Planet, One Future*. I also extend my personal appreciation to Rebecca Brennan and Leila Walker at Taylor & Francis (Routledge) for their patience and guidance in developing the book, along with Anne Elliott-Day for her skilful copy-editing on behalf of Wearset.

Many individuals and organisations (*italics*) have contributed to this work by reviewing chapters, providing text, sharing insights, facilitating copyright permission, helping with illustrations or offering guidance and encouragement.

In particular, I thank Michael Abramowitz, Amalia Almada, Badr Al-Olama, Karen Alroy, *American Public Health Association (APHA)*, Kofi Annan, Bryan Appleyard, *Association of Schools of Public Health in the European Region (ASPHER)*, Astrid Auraldsson, David Attenborough, Sara Babo-Martins, *BBC News*, *BBC Radio 4*, Lars Barteit, Casey Barton Behravesh, Marielle Bemelmans, Franck Berthe, Veronica Boix Mansille, Isabelle Bolon, Michael Boskin, Timothy Bouley, Tianna Brand, *British Council*, *Brookings Institute Centre for Health Policy*, Edmund Brown, Gordon Brown, Alex Brummer, *Bulletin of the Atomic Scientists*, Michael Burleigh, Lisa Buren, Genc Burazeri, Tanya Byron, *Carbon Literacy Project*, Sara Carr, Gerardo Ceballos, *Centers for Disease Control and Prevention (CDC)*, *Chatham House* (London), *Chartered Global Management Accountant*, *Civil Society Futures*, *Club of Rome*, Tracy Collins, Kristin Cohen, *Commonwealth Secretariat*, Loren Cordain, Kasia Czabanowska, Janet Daley, Stephen Dorey, David de Pooter, *Department of Peace and Conflict Research* (Uppsala University), Jared Diamond, Liesel Ebersohn, *EcoHealth Alliance*, *EcoWatch*, Jeanette Elsworth, Tom Engelhardt, *European Council on Foreign Relations*, Jimmy Fallon, Natalie Fenton, Niall Ferguson, Helen Filby, Gordana Filipic, Antoine Flahault, Martin Fletcher, Lawrence

Freedman, *Freedom House*, *Frontier Strategy Group*, Jane Ganeau, Rainer Gerling, Annette Gerritsen, *Global Partnership for Education*, *Global Policy Forum*, James Gifford, Christopher Golden, Olivia Goldhill, Sandy Grande, Gregory Gray, António Guterres, Barbara Haesler, Andrew Haines, Yuval Harari, Jessica Hart, Jolene Hattingh, Stephen Hawking, Jeffrey Holland, Bethany Hughes, *Institute for Economics and Peace*, *Institute of Global Health* (University of Geneva), *International Council for Science*, *International Development Research Centre*, *Institute for Health and Aging and Department of Social and Behavioural Sciences* (University of California), *InterAction Council*, *International Veterinary Students' Association (IVSA)*, *International Physicians for the Prevention of Nuclear War*, *Japan Medical Association*, *Japan Veterinary Association*, Charles (Chuck) Johnson, William Joy, Nazir Kazmi, Ban Ki-moon, Ashok Khosla, Jim Yon Kim, Lonnie King, Rita J King, Tomiko Hokama, Otmar Kloiber Keeley Knowles, Donu Kogbara, Phil Korbel, Rosina Krecek, Vipat Kuruchittham, Mordecai Kurz, Ray Kurzweil, Miroslav Lajčák, Emma Lefton, Alexander Likhotal, Anthony Lloyd, Natalia Linou, Megan Linquiti, Jill Lueddeke, Pnina Luttinger, Brian Lutz, Catherine Lutz, Ellen MacArthur, Catherine Machalaba, Ellen MacArthur, John Mackey, Emmanuel Macron, Selaelo Mametja, Tshepo Matjila, John McKenzie, *McKinsey Global Institute*, Leila Mead, Mary Meehan, Angela Merkel, Hannah Metz, Rana Mitter, Oliver Moody, Bernard Mutsago, Samuel Myers, Jordan Naidoo, *National Center for Emerging and Zoonotic Infectious Diseases*, *National Institute for Child and Human Development*, *National Intelligence Council*, *Network for Evaluation of One Health (NEOH)*, Zeev Noga, Johan Norberg, Joanna Nurse, Barack Obama, Martin O'Halloran, Jim O'Neill, *One Health Platform (OHP)*, *One Health Tripartite Collaboration (FAO, OIE and WHO)*, Organisation for Steven Osofsky, Ab Osterhaus, Homer Wilson, John Wyn Owen, Yuhyun Park, Kyle Pearce, *Peter G Peterson Foundation*, Therese Pettersson, Thomas Piketty, Steven Pinker, *Planetary Health Alliance (PHA)*, Mit Philips, Jonathan Powell, *PricewaterhouseCoopers Project Initiatives*, Vladimir Putin, Kevin Queenan, Peter Rabinowitz, Prince Harry, Afifah Rahman-Shepherd, Simon Rattle, Lord Rees, Matt Riddley, George Ritzer, Johan Rockström, Deborah Rogers, Justin Rosenstein, Robert Ruby, Simon Rüegg, Rafael Ruiz de Castañeda, Helen Rumbelow, Andreas Schleicher, Klaus Schwab, Richard Seifman, Flavia Senkubuge, Danny Sheath, Mike Shooter, Rohan Silva, Julie Sinclair, Raj Sisodia, Pindai Sithole, *sparks & honey*, *Southeast Asia One Health University Network (SEAOHUN)*, Julia Spencer, *Statistics Times*, Achim Steiner, *Stockholm Resilience Centre*, Matthew Stone, Alexandru Supeanu, Matthew Symond, Kaiser Tarafdar, *techUK*, *TED Talk*, *Transparency International*, Gregory Treverton, Farooq Ullah, *UN Department of Economic and Social Affairs*, *United Nations Development Programme (UNDP)*, *UN Development Statistics and Information Branch*, Lao Tzu, *UN Habitat*, *UN Industrial Development Organization*, *UN News*, Julia Unwin, Denise van Eekelin, Chris Vanlangendonck, Daniel Wahl, David Waltner-Toews, Helen Wang, *Watson Institute Costs of War Project* (Brown University), Ruby Wax, Cleo Wade, Helen Wang, Douglas Webb, Justin Welby, Joel Whitaker, Roger Williamson, A. Lee Willingham, Dorothy Wills, *World Bank Group (WBG)*, *World Economic Forum*, *World Obesity Federation*, *World Medical Association (WMA*, *World Veterinary Association (WVA)*, *World Wildlife Fund (WWF)*, Judith Woods, *Worldodometers*, Malala Yousafzai, Muhammad Yunus, Joe Zarif.

Introduction

The state of planet Earth: time to 'stop and think'!

We carry within us the wonders we seek without us.

Sir Thomas Browne (1605–1682)

The bleak possibility of a cold technology-driven algorithmic world contrasts sharply with the heartfelt and lone voice of a young Irish girl who was invited to talk about her generation's future 'hopes and dreams' at the 2013 G8 summit in Belfast, Ireland.[1] Before a keynote speech delivered by Barack Obama, then president of the United States, she conveyed a simple yet compelling message to world leaders – all she wanted was 'to live in a society where we are safe and can be friends with everybody'.

Given today's global uncertainties, it is clear that her plea 'to make the world a fairer and better place' is now further from reality as we continue to face urgent and interconnected problems on many social, economic, geopolitical and environmental fronts.[2] Many of these have come under the spotlight only in the past few years, including planetary boundaries (environmental limits), resource shortages, ideological extremism, cyberwarfare (destabilising the foundations of Western democracy) and the threat of nuclear war by 'nuke' or rogue states.

Arguably, some of these threats could have been avoided if the world had paid more attention to, and enacted, the principles of the United Nations *Earth Charter*, launched on 29 June 2000.[3] This comprehensive ethical framework recognised the need to 'join together to bring forth a sustainable global society founded on respect for nature, universal human rights, economic justice, and a culture of peace'.

Now endorsed by over 6,000 organisations, including many governments and international organisations, the Charter's key principles are as relevant as ever – indeed, as cautioned below, more so:

- Respect and Care for the Community of Life
- Ecological Integrity
- Social and Economic Justice
- Democracy, Nonviolence and Peace.

With the probability of global catastrophe rated as 'very high', the Doomsday Clock, maintained by members of the *Bulletin of the Atomic Scientists*' Science and Security Board,[4] has now been moved closer to midnight, from two and a half minutes to just two minutes. The urgency of addressing the geopolitical challenges that inform the scientists is undeniable.

Equally crucial for global sustainability are 'human impacts on natural systems'. At the 2017 US Academy of Medical Sciences and *The Lancet* International Health Lecture, Professor Sam Myers of Harvard University succinctly and irrefutably summarised the state of the planet:[5]

> Driven by rapid increases in the size of the human population and even steeper growth in per-capita consumption, the scale of human impacts on our planet's natural systems is hard to overstate: to feed ourselves, we annually appropriate about 40% of the ice-free, desert free terrestrial surface for pastures and croplands; we use about half of the planet's accessible water, largely to irrigate our crops, and we exploit 90% of global fisheries at, or beyond, their maximum sustainable limits. In the process, we have cut down 7–11 million km^2 of the world's forests and dammed more than 60% of its rivers. The quality of air, water, and land is diminishing in many parts of the world because of increasing global pollution. These and other processes are driving species to extinction at roughly 1000 times baseline rates while reducing population sizes of mammals, fishes, birds, reptiles, and amphibians by half in the past 45 years.

The One Health Tripartite,[6] involving the World Health Organization (WHO), the World Organisation for Animal Health (OIE), and the UN Food and Agriculture Organization (FAO) has convincingly argued the importance of the need to take 'a more integrated approach to health'; this has also been echoed by the authors of a seminal paper on human, animal, plant and ecosystem health.[7]

Along with OIE, WHO and FAO, leading global *One Health* and related advocates include the Centers for Disease Control and Prevention (CDC), Chatham House, the EcoHealth Alliance, the InterAction Council, the International Veterinary Students' Association, the Institute for Global Health, the Network for the Evaluation of One Health, the One Health Commission (OHC), the One Health Initiative (OHI), the One Health Platform, the Planetary Health Alliance, the Southeast Asia One Health University Network, the United Nations Development Programme, the World Bank, the World Federation of Public Health Associations, the World Medical Association (WMA), the World Veterinary Association (WVA), and the World Wildlife Fund for Nature (WWF) International (see Chapter 11).

Together, these and other organisations and movements – in cooperation with governments and civil society – have the potential to help prevent a potential dystopic planet and shift policy priorities towards ensuring 'our responsibility to one another' and 'to the greater community of life'.[3] Finding collaborative solutions to transform our world and 'to end poverty, protect the planet and ensure that all people enjoy peace and prosperity'[8] must surely be shifted to the top of all government and corporate agendas as global threats continue unabated.

Collectively, the aforementioned organisations – joined by others with like-minded concerns – could also take a lead in finding the global consensus sought by William Joy, an American computer scientist, several years ago.[9] His plea to agree '*as a species, what we wanted, where we are headed, and why* [italics added]'[9] is as crucial now as it was then – perhaps much more so. Furthermore, he concluded that if we are successful in achieving this global harmony 'we could make our future much less dangerous' and 'then we might understand what we can and should relinquish'. Following Joy's aspirations, we might even be able 'to start,' as Professor Harari urges, to look 'towards new horizons'.[10]

The realisation that the human species is deeply embedded in the natural world and the survival of all species, including our own, is wholly dependent on a healthy planet is becoming increasingly recognised in the first few decades of the twenty-first century. Words, however, do not always translate into action and the health of our planet continues to be in serious trouble. According to the authors of 'Safeguarding human health in the Anthropocene epoch: Report of The Rockefeller Foundation–Lancet Commission on planetary health',[11] 'human health is better today than any time in human history' as life expectancy has risen across the globe and child mortality rates have fallen significantly. However, our progress to date has come at a heavy price – some have called it the 'ecology paradox' – as 'we have been mortgaging the health of future generations in order to realize economic and development gains in the present.'[11] Evidence of this *trade-off* abounds, including, as highlighted by the Center for One Health Research at the University of Washington, 'biodiversity loss, climate change, particulate air pollution, ocean acidification, and deforestation.'[12]

A specific example that confirms 'the degradation of nature's life support systems,' highlighted previously,[5] is cited in *The Living Planet Index*[13] and should 'make us stop and think':

> In less than two human generations, population sizes of vertebrate species have dropped by half. These are the living forms that constitute the fabric of the ecosystems which sustain life on earth and the barometer of what we are doing to our planet, our only home.[13]

The Rockefeller report underscores how 'this environmental degradation can pose a serious threat to human health in the future, and calls for urgent research and policy action to address these large-scale problems',[11] particularly considering past failures.

It is therefore reassuring that global initiatives to mitigate the effects of an increasingly compromised natural world – including the very real potential of ecosystem collapse – are taking place not only through transdisciplinary scientific research but also on the social front, as governments are beginning to realise that policies must 'balance social progress, environmental sustainability and the economy'.[14] These are closely complemented by educational, research and information-sharing initiatives driven by such organisations as the WVA–WMA alongside the Japan Medical Association–Japan Veterinary Association.[15]

Public pronouncements, social media, books and articles by prominent global spiritual leaders and thinkers – such as Pope Francis, the late physicist Professor Stephen Hawking, the conservationist Sir David Attenborough and African social change advocates, Dr Joyce Nyairo, John Mbaria and Dr Mordecai Ogada, to name a few – have been intensifying, collectively voicing the 'folly of a limitless world'.[16] The urgency to confront 'giants with big funding, guns and deep international networks'[17] is particularly acute in sub-Saharan Africa nations.

Taken together, their message is clear: the human race is highly vulnerable to exploitation and profiteering resulting in global warming, inequalities, inequities, corruption, economic uncertainties, ideological extremism, food insecurity and the potential for global conflict. The status quo is unsustainable and we must not only do things better but also do better things.[18]

Ways forward apply as much to economic and political developments as they do to the environment. In terms of the former, Niall Ferguson, a British historian at Oxford University as well as at Stanford's Hoover Institution and a commentator, reflects on the future and sees parallels between the 'Anglo-German antagonism' before World War I and today's tensions between America and China.[19] The choice for both superpowers, he says, lies between 'a trade war and escalating into a real one' and accepting, as former US Secretary of State, Dr Henry Kissinger, advised in his book *On China*,[20] the principle of 'co-evolution'. In his concluding paragraphs, Kissinger highlights that 'the two countries owe it to themselves and the world to attempt to do so.' Moreover, 'The practice of coevolution,' he cautions, 'will take wisdom and patience. After some time, it may grow into an integral part of each country's conduct.'[20]

In her book, *The Chinese Dream*,[21] business consultant and China expert, Helen Wang, reinforces this approach and underscores that 'co-evolution' can generate 'a *sense* of common purpose', based on recognising 'different ways of thinking'. The future, she envisions, is one that ensures 'the United States and China pool efforts, with each other and with other states, to bring about an agreed world order'. Placing the criticality of succeeding in this regard in perspective, Bryan Appleyard, British journalist and author, concludes that after millions of years of evolution, we are, after all, only 'temporary arrangements of biological matter in the midst of an ocean of living creatures, whose depths we have just begun to imagine'.[22]

Preventing us from respecting *all* life on the planet – including our own species, *Homo sapiens* – is our delusionary sense of power. Arguably, it could be this innate human character flaw (the Achilles' heel of our civilisations?) that is the root cause of most of the global problems we face – especially the assumption that 'more is better'.[18] For Sir David Attenborough, the esteemed veteran British naturalist, this self-delusion has led to a situation, to cite one example, where 'the world's oceans' are turning 'into a "toxic soup" of industrial waste and plastic, putting the future of humanity at risk'.[23]

> The fact is that no species has ever had such wholesale control over everything on earth, living or dead, as we now have. That lies upon us, whether we like it or not, an awesome responsibility. In our hands now lies not only our own future, but that of all other living creatures with whom we share the earth.[24]

'Knowing what we know', all of us are now challenged 'to stop and think' about the sustainability of planet Earth and the 'sanctity of life'[25] – the *survival* of all species. As former UN Secretary-General Ban Ki-moon made clear in 2014: 'there is no "Plan B" for action, as there is no "Planet B"'.[25]

Writing for the Carbon Literacy Project[26], a 'world-first initiative in its aims to empower the citizens of Manchester (UK) and beyond with Carbon Literacy', project officer Helen Filby provides an insightful and compelling argument in support of the *One Health* concept and why we should take the words of 'the late, great physicist Stephen Hawking' very seriously:

> Personally, I think it makes much more sense to first try everything possible to protect this sure world that we live on from irreversible damage, encompassing the environment, ecosystem, as well as all animals and plants that live here; rather than chancing that some yet to be discovered science might save us. Technology has already advanced enough that we have many options available to live a more sustainable life on this planet, most notable with the multitude of renewable energies available. Continued education among generations will help to combat the lack of willingness to change. When a push for change comes from within society itself, it is often more likely to be effective because it is not being forced upon the majority; it is a choice, like for example the current push to reduce plastic pollution stemming from the educational BBC series 'Blue Planet II'. It is important for governments, businesses and organisations the world over to engage with climate change, putting long term sustainability in front of short term profits. Saving ourselves from avoidable man-made environmental damage is still within reach, but only by working together as a global community can we hope to create a promising future on this beautiful planet for all, and most importantly for future generations.[27]

References

1 Spencer BJ. *Hannah Nelson speech in full* [and video]. http://brianjohnspencer.blogspot.co.uk/2013/06/hannah-nelson-speech-in-full.html (accessed 14 September 2017).
2 Lueddeke GR. *Global Population Health and Well-Being in the 21st Century: Toward New Paradigms, Policy and Practice*. New York: Springer Publications; 2016.
3 Earth Charter International. *The earth charter*. http://earthcharter.org/invent/images/uploads/echarter_english.pdf (accessed 28 January 2018).

4 Timeline – It is two minutes to midnight. *Bulletin of the Atomic Scientists*. https://thebulletin.org/timeline (accessed 30 January 2018).
5 Myers SS. Planetary health: Protecting human health on a rapidly changing planet. *The Lancet* 2017; 390(10114): 2860–2868.
6 *FAO–OIE–WHO collaboration concept note on health risks at the human–animal interface*. www.oie.int/for-the-media/press-releases/detail/article/fao-oie-who-collaboration-concept-note-on-health-risks-at-the-human-animal-interface-is-available-on/ (accessed 27 September 2017).
7 Queenan K, Garnier J, Nielsen RL, Buttigieg S, de Meneghi D, Holmberg M, *et al. Roadmap to a One Health agenda 2030*. www.cabi.org/CABReviews/FullTextPDF/2017/20173134856.pdf (accessed 25 November 2017).
8 United Nations Development Programme. *What are the Sustainable Development Goals?* www.undp.org/content/undp/en/home/sustainable-development-goals.html (accessed 20 January 2018).
9 Joy W. *Why the future doesn't need us*. www.wired.com/2000/04/joy-2/ (accessed 27 September 2017).
10 Harari YN. *Homo Deus: A Brief History of Tomorrow*. London: Harvill Secker (Vintage); 2016.
11 Whitmee S, Haines A, Beyrer C, Boltz F, Capon AG, De Souza Dias, BF, *et al*. Safeguarding human health in the Anthropocene epoch: Report of The Rockefeller Foundation–Lancet Commission on planetary health. *The Lancet* 2015; 386(1007): 1973–2028.
12 Rabinowitz P. *One health and planetary health*. http://deohs.washington.edu/cohr/blog/2017/15/one-health-and-planetary-health (accessed 14 November 2017).
13 Foreword – message from WWF International Director General. *Living planet report 2014 – Species and spaces, people and places 2014*. http://assets.worldwildlife.org/publications/723/files/original/WWF-LPR2014-low_res.pdf?1413912230 (accessed 29 May 2016).
14 United Nations. *Balancing social, environmental and economic considerations in procurement – Supplement to the 2012 annual statistical report on United Nations procurement*. www.ungm.org/Areas/Public/Downloads/2012_Annual_Statistical_Report_supplement.pdf (accessed 18 September 2017).
15 The 2nd World Veterinary Association–World Medical Association Global Conference on One Health: Moving forward from One Health concept to One Health approach, Kitakyushu City, Fukuoka Prefecture, 10–11 November 2016. *Japan Medical Association Journal* 2016; 59(4): 149–153. www.ncbi.nlm.nih.gov/pmc/articles/PMC5476991/ (accessed 4 July 2018).
16 One Health Commission. *What is One Health?* www.onehealthcommission.org/en/why_one_health/what_is_one_health/ (accessed 10 October 2017).
17 Mbaria J, Ogada M. *The Big Conservation Lie*. Auburn, WA: Lens and Pens Publishing; 2017.
18 Lueddeke G. Learning not only 'to do things better' but also 'to do better things'. *Africa Health News*, 28 September 2017. https://africahealthnews.com/learning-not-things-better-also-better-things/ (accessed 18 January 2018).
19 Ferguson N. As Trump and Xi spar, an ancient trap awaits them both. *The Sunday Times*, 6 May 2018: 23.
20 Kissinger H. *On China*. New York: Penguin Press; 2012.
21 Wang H. *The Chinese Dream: The Rise of the World's Largest Middle Class and What It Means to You*. Charleston, SC: Createspace; 2010.
22 Appleyard B. Dirt is your friend. *The Sunday Times Magazine*, 15 January 2017.
23 Knapton, S. Sir David Attenborough: The world's oceans are becoming a 'toxic soup' of industrial waste and plastic. *Telegraph*, 5 December 2017.
24 Professor Ailsa Hall. Graduation address, 21 June 2017. www.st-andrews.ac.uk/graduation/previous-graduation/graduation-2017/graduation-addresses/ailsa-hall/ (accessed 4 July 2018).

25 Mohan V. *There is no 'Plan B' for action, as there is no 'Planet B'*. https://timesofindia.indiatimes.com/world/us/There-is-no-plan-B-for-action-as-there-is-no-planet-B-UN-chief-says-ahead-of-UN-Climate-Summit/articleshow/43134461.cms (accessed 25 February 2017).
26 Carbon Literacy Project. *Home*. www.carbonliteracy.com/ (accessed 25 March 2018).
27 Filby H. *Stephen Hawking's outlook on climate change*. www.carbonliteracy.com/stephen-hawkings-outlook-on-climate-change/ (accessed 20 March 2018).

Part I

Safeguarding the health and well-being of people and planet

1 Threats to the health and sustainability of populations

From rural to urban life

Ensuring the future sustainability of life on Earth is becoming much more problematical as threats such as climate change, the loss of biological diversity, water scarcity and food insecurity are growing at unprecedented rates and scale. The convergence of these and other risks are being exacerbated considerably as the global population shifts from largely agrarian to urban societies.

As shown in Figure 1.1, the UN Department of Economic and Social Affairs graphs illustrate that the most urbanised regions include Northern America (82 per cent living in urban areas in 2014), Latin America and the Caribbean (80 per cent in urban areas), and Europe (73 per cent in urban areas). In contrast, Africa and Asia remain mostly rural, with 40 per cent and 48 per cent of their respective populations living in urban areas.[1]

It is estimated that by 2050 more than 70 per cent of the world's population – about 9.3 billion (compared with around 7.6 billion today) will be city-dwellers.[2] City population increases will be greatest in Africa, the Middle East, India, China and southeast Asia, including increases in the number of megacities with total populations of over 10 million people, such as Tokyo and Mumbai.[3] Currently, there are 25 megacities across the world and this will rise to 35 by 2025 with some of the fastest growth occurring in Dhaka (population increase of around 53 per cent), Beijing (expanding by 44 per cent), New Delhi (population rising by 43 per cent) and Tokyo (population increasing from 37 million to 39 million).[3]

The African continent

In 1980, only 28 per cent of Africans lived in urban areas; today, approximately 40 per cent have settled in towns and cities. In total, the region's population reached 1 billion in 2009 and joined India and China as the third most populated region of the world.[4]

Figure 1.2 compares today's population in Africa's largest cities to expected growth by 2025 and shows that many large cities, such as Dakar, Kinshasa and Luanda, will almost double in size 2025[5] Overall, Africa's population is expected to grow by more than 500 million people in the next few decades and a further

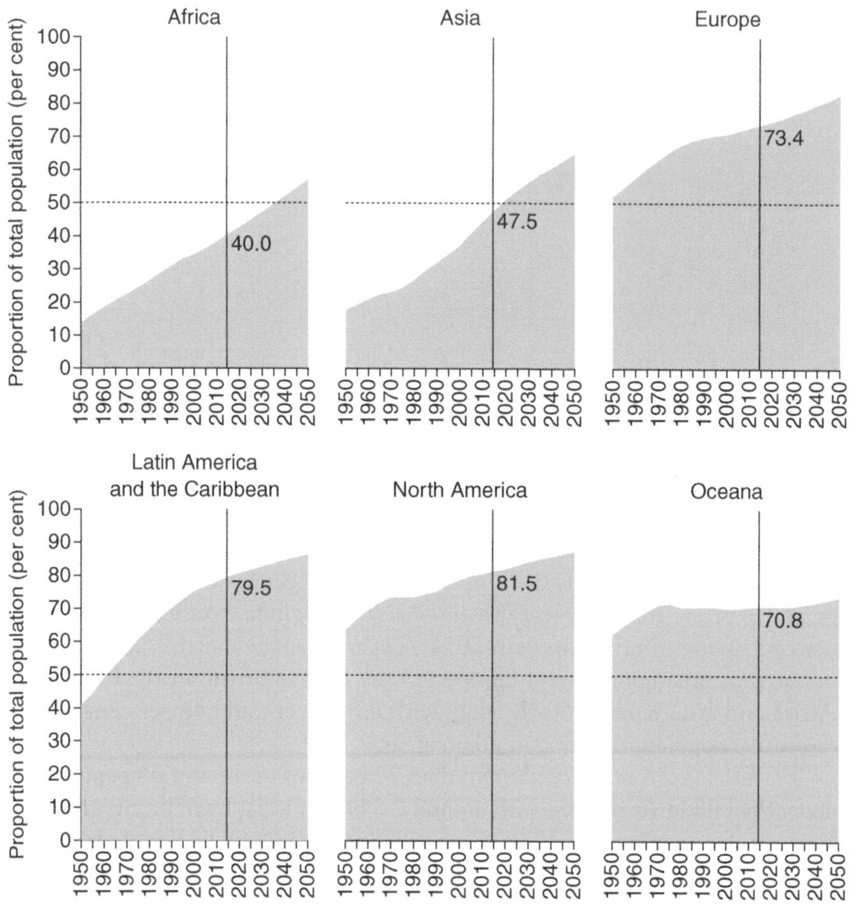

Figure 1.1 Urban and rural population as proportion of total population, by major areas, 1950–2050.

Source: *World urbanization prospects*. Department of Economic and Social Affairs, © 2014 United Nations. Reprinted with the permission of the United Nations.

500 million by 2050, with 60 per cent of its population living in cities, compared to 1950 when the continent had fewer than 500,000 urban dwellers.

By 2100 it is predicted that Africa's population will be between 9 and 13 billion. Given that currently about 70 per cent of Africa's urban population live in slum conditions, the challenge to already 'overstretched and under-resourced municipal authorities to provide water, electricity and social services, not to mention governance and security'[5] will continue to be immense and will place additional pressure on governments to find ways of managing accelerating rural–urban migration (largely unregulated) in a way that ensures equity.

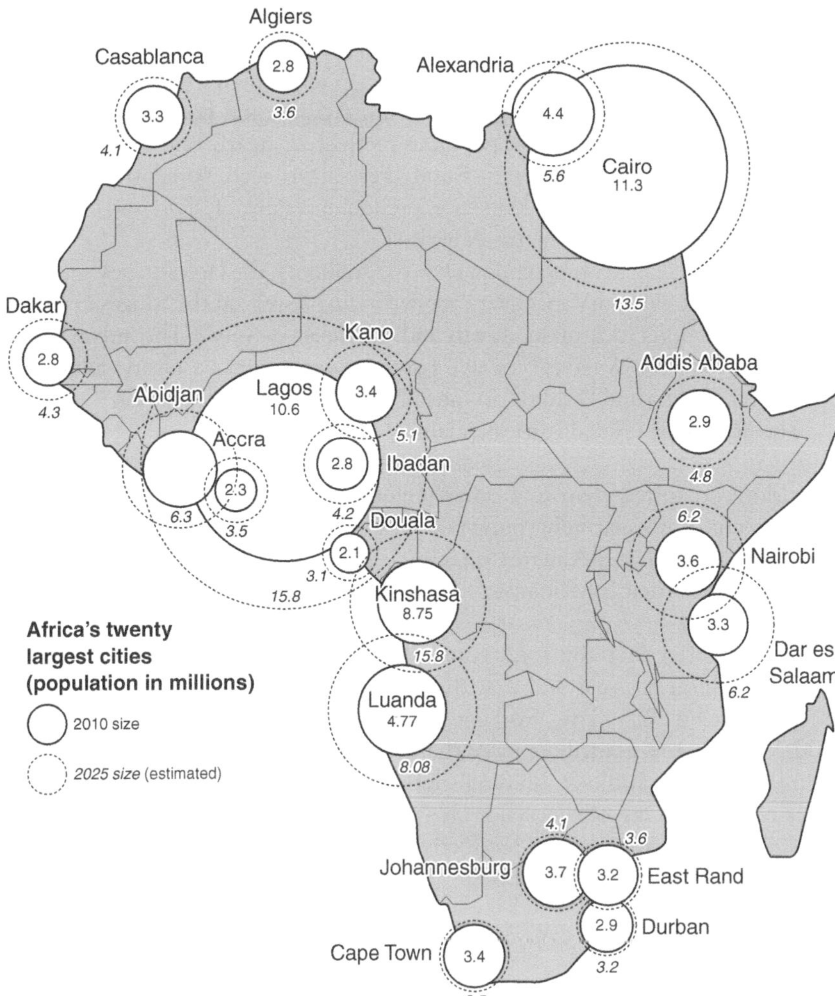

Figure 1.2 Growth of Africa's largest cities.

Source: Frontier Strategy Group, *Capturing Africa's transformative urban growth* (2011).

The Americas

While overall urban population densities in the Americas will remain generally stable to 2050, averaging around 80 per cent, several cities will experience significant growth by 2025, including Mexico City (20 to 25 million – 25 per cent growth), New York (20 to 24 million – 20 per cent growth), Los Angeles (13 to 16 million – 23 per cent growth), Rio de Janeiro (12 to 14 million –17 per cent growth) and Sao Paulo (20 to 23 million –15 per cent growth).[6]

In the US, with a population of around 327 million in 2018 (4.28 per cent of the world's population, increasing by around 2 million annually),[7] 'Nearly one in seven Americans lives in the metropolitan areas of the country's three largest cities: New York, Los Angeles and Chicago.'[8] Adding 'metro Dallas-Fort Worth, Houston, Philadelphia, Washington, Miami, Atlanta and Boston to that group (totaling the country's 10 most populous metros) ... nearly one in three Americans lives in these few spots'.[8] Rural areas have seen population declines, 'neither attracting new residents nor producing many of their own' with an ageing population in 'non-metro America'.[8]

Similarly, in Canada (population about 37 million[7]), the largest cities, including Toronto, Montreal and Vancouver, 'are expanding fastest on the fringes in spite of government efforts to contain growth and encourage density'.[9] The 'urban spread' is towards Canada's West as 'The five fastest-growing cities – Calgary, Edmonton, Saskatoon, Regina and Lethbridge – are all in Alberta or Saskatchewan'.[9]

The McKinsey Global Institute claims:

> More urbanized than any other region in the developing world, with 80 percent of its relatively young population (total about 428m, including the Caribbean), Latin America is feeling the strain in terms of delivering basic services to their inhabitants.[10]

It is anticipated that the total population of Latin America will increase by about 4 million per year and that the number living in cities will rise to about 85 per cent by 2025. The 198 largest cities – defined as having populations of 200,000 or more – contribute more than 60 per cent of GDP but many 'are grappling with traffic gridlock, housing shortages, and other symptoms of diseconomies of scale'.[10] One of the major challenges facing the region's top 10 cities is 'to significantly enhance the productivity and number of jobs they generate in the formal economy and boost the efficiency of their operations and management'.[10]

Without decisive action by political and business leaders, it is possible that

> By the second half of this century, Latin America's demographic profile will look more like Europe's, with a shrinking proportion of economically active young having to provide for a growing fraction of older people. The risk is that the region could grow old before it becomes rich if cities do not fulfill their potential.[10]

Statistics by the United Nations (UN) 'show that the Caribbean (c.70m) is one of the most highly urbanised regions in the world, with 66.2 percent of its population living in urban settlements – a proportion almost twice as high as those for Africa and Asia (39%)'.[11] However, 'Since urbanization and urban planning are not viewed as crucial issues, the consequence has been ineffective land use, culminating in the avoidable loss of valuable lands which could be used for other pressing environmental and social interventions.'[11]

According to Stacey Thomas at the University of the West Indies in Trinidad,[11] 'The multitude of pressing problems that confront the region – such as informal settlements, environmental degradation, resource exhaustion, and underdeveloped infrastructure – necessitate immediate attention by planners.' However, advocating 'any type of sustainable urban form, compact or otherwise' is difficult given 'limited supporting documentation and a general lack of research on urban areas'.[11] It is clear that 'urbanization and urban planning are not viewed as crucial issues'.[11] As a consequence, there has been 'ineffective land use, culminating in the avoidable loss of valuable lands which could be used for other pressing environmental and social interventions'.[11]

China: managing expanding populations

Unlike the estimated lower population growth in the Americas, China's population – about 1.5 billion in 2018, increasing by about 6 million each year[7] – will likely grow by more than 500 million in the next 17 years and a further 500 million by 2050,[12] compared to 1950 when China had fewer than 500,000 urban dwellers. By 2050 it is expected that the nation will have 221 cities with more than 1 million inhabitants of which 23 cities will have more than 5 million people and 8 megacities will have populations of more than 10 million. While distinct threats to infrastructure capacity are already being felt as over 200,000 people are presently moving into cities every day – 72.8 million people annually – there may also be considerable advantages.

Citing several examples, McKinsey Global Institute research

> suggests that concentrated urban growth scenarios could produce 20 percent higher per capita GDP than that yielded by China's current urbanization path, would have higher energy consumption but also higher energy efficiency, and would contain the loss of arable land.[12]

In addition, from a wealth-generating perspective, 'Concentrated urbanization would also have the advantage of clustering the most skilled workers in urban centers that would be engines of economic growth, enabling China to move more rapidly to higher-value-added activities.'[12]

Nevertheless, major pressures on the economy will be considerable for many cities 'including the challenge of managing these expanding populations, securing sufficient public funding for the provision of social services, and dealing with demand and supply pressures on land, energy, water, and the environment.'[13]

India: public health challenges

Similarly, India's rapid population growth (total around 1.4 billion in 2018,[7] increasing by about 5 million per year) is exemplified by a new model city, Gurgaon. In just 20 years this 'sleepy town surrounded by miles of open farmland',[13] only 12 miles south of Delhi, has now grown to a city of about 17 million people, increasing by around 1 million annually, thereby becoming India's fastest growing city.[13] The city boasts 'skyscrapers, shopping malls, and gated residential developments, a corporate park called "Cyber City" housing some of the world's leading IT and commercial companies'.[13] And, while outwardly a symbol of progress, it now suffers from very serious public health challenges: 'chronic water shortages, a power supply so erratic that business and private residents are obliged to provide their own backup generators and a drainage system so inadequate that in monsoon time sewage bubbles up on the pavements.'[13]

Urbanisation as 'constructive disruption'

Barry Wilson, who heads 'an innovative, international construction design and management consultancy'[14] in Hong Kong, cites the UN Department of Economic and Social Affairs 2014 report, 'World urbanization prospects,[2] and summarises the advantages and disadvantages of urbanisation.[14] The benefits, he says, include:

- expanding 'access to the key services of education and healthcare for large numbers of people in an economically efficient manner';
- providing 'public transportation, as well as housing, electricity, water and sanitation for a densely settled population is typically cheaper and less environmentally damaging than providing a similar level of services to a predominantly rural household'; and
- accessing 'larger and more diversified labour markets'.[14]

In contrast, Wilson underscores the disadvantages: 'the detachment from family and friends … initial sense of alienation; high cost of living, pollution and stress; high risk of epidemics, natural disaster and social disruption given their concentrations of population and infrastructure'; along with 'their inherent instabilities and vulnerabilities' and a number of other problems.[14]

Giving a TEDx talk in China on 'Future city planning learned from science fiction movies',[15] or constructive disruption, Wilson's main message was that:

> Our cities and lifestyles have been shaped by what's gone before. In the midst of what is now the 4th industrial revolution, as the impossible becomes possible, we urgently need to relinquish outdated 'planning as usual models', envision the world in which we want to live, and then urgently take steps to change it and shape it for the better. Time is against us.

In a speech at the promotion conference of high-tech parks in Shenzhen, Heyuan, Shanwei and Guang'an,[16] Wilson mentioned urban developments in the UK over the past 50 years and progress made in cities such as Manchester, Birmingham and Liverpool. He concluded that: 'Education was the key to a new generation of thinking about the environment we wanted to live in. Change in behavior was essential.'[16] However, Wilson also emphasised that the global challenges we face today are even more difficult than those experienced by Britain in the 1970s and beyond, including 'maintaining the huge and aging populations moving into cities', through urbanisation as well as the 'risk of major and repeated catastrophes caused by climate change … the loss of cultural heritage and diversity from globalization [and] environmental degradation from developing in the wrong place'.[16]

Wilson's fundamental guiding principle at the China conference was the 'need to plan and design by anticipating tomorrow's world. By expecting the unexpected. By envisioning our own futures'.[16] In particular, he anticipated new lifestyles based on three key areas: future transport and technology needs; minimised environmental, social and cultural impacts; and reduced disaster risk. Box 1.1 provides an illustration of the future urban life that Wilson envisages, integrating the traditional values of family life augmented by environmental considerations and advances in technology.[16]

Box 1.1 Imagining future urban life

'I'm looking out of the window of my home. It used to be an old glass factory but was converted into a high-tech park and mixed development of affordable housing, elderly accommodation and loft accommodation for young people. It's really friendly and has lots of activities. There is a kindergarden and a healthcare centre where the staff check on the local residents using both IT and regular personal visits. It has a business incubator where I run my new "city planning" business so I don't have to travel to work every day.

Outside the window I am surrounded by trees and see a stream with white ducks on it. I can see far into the distance and the sky is bright blue. There are no roads and I see people leaving to go to their work from the rooftop aerial taxi station and others in electric pods, slowly mingling with bikes and pedestrians on their way to the metro station. There is a "ring ring" and an image of my daughter pops up as a hologram in front of me. She tells me it's a beautiful day outside and suggests I join her at the local community farm on the roof to do some gardening and eat our own grown vegetables together for lunch.'[16]

While not everyone may share Wilson's particular vision of an urban 'high-tech park' lifestyle, most could perhaps agree that the question, 'What future do you want to build for your children?' should be at the centre of all prospective planning and construction scenarios.

References

1 United Nations. *World's population increasingly urban with more than half living in urban areas*. https://esa.un.org/unpd/wup/publications/files/wup2014-highlights.Pdf. (accessed 27 September 2017).

2 United Nations, Department of Economic and Social Affairs. *World urbanization prospects*. www.un.org/en/development/desa/publications/2014-revision-world-urbanization-prospects.html (accessed 9 July 2018).

3 McCarthy N. *The world's megacities are set for major growth*. www.statista.com/chart/1826/population-growth-in-the-worlds-megacities/ (accessed 25 June 2017).
4 Population Connection. *African urbanization*. www.populationconnection.org/article/african-urbanization/ (accessed 12 February).
5 Frontier Strategy Group. *Capturing Africa's transformative urban growth*. http://blog.frontierstrategygroup.com/2011/08/capturing-africa%E2%80%99s-transformative-urban-growth/ (accessed 9 July 2018).
6 United Nations, Department of Economic and Social Affairs. *World urbanization prospects – The 2014 revision*. https://esa.un.org/unpd/wup/publications/files/wup2014-highlights.pdf (accessed 9 July 2018).
7 Worldodometers. *Current world population*. www.worldometers.info/world-population/ (accessed 19 March 2018).
8 Badger E. *Metropolitan areas are now fueling virtually all of America's population growth*. www.washingtonpost.com/news/wonk/wp/2014/03/27/metropolitan-areas-are-now-fueling-virtually-all-of-americas-population-growth/?utm_term=.b1bf7cf6c95b (accessed 22 November 2017).
9 Curry E. *Big Canadian cities see faster suburban growth despite bid to boost density*. www.theglobeandmail.com/news/national/census-2016-big-canadian-cities-see-urban-spread-but-rural-communities-wane/article33949353/ (accessed 18 November 2017).
10 McKinsey Global Institute. *Fulfilling the promise of Latin America's cities*. www.mckinsey.com/global-themes/urbanization/fulfilling-the-promise-of-latin-americas-cities (accessed 27 October 2017).
11 Thomas S. *'The Future is Urban' – Challenge of sustainable urban development in the Caribbean: The search for sustainable urban forms*. www.isocarp.net/Data/case_studies/2328.pdf (accessed 9 July 2018).
12 Woetzel J, Mendonca L, Devan J, Negri S, Hu Y, Jordan L, *et al*. *Preparing for China's urban billion*. www.mckinsey.com/global-themes/urbanization/preparing-for-chinas-urban-billion (accessed 18 September 2017).
13 Brown M. Growing pains. *Telegraph*, 25 May 2013: 24–33.
14 Barry Wilson Project Initiatives. *Urban agglomeration a high risk strategy*. www.initiatives.com.hk/150326-urban-agglomeration (accessed 21 June 2018).
15 Wilson B. *Future city planning learned from science fiction movies (constructive disruption)* [video]. http://v.youku.com/v_show/id_XMjQ4OTgzMzYyNA==.html (accessed 21 June 2018).
16 Wilson B. *Futuristic urban solutions will take bold vision*. www.scmp.com/comment/insight-opinion/article/2102289/futuristic-urban-solutions-will-take-bold-vision (accessed 21 June 2018).

2 Climate change

Focus on Africa, Asia and the coastal poor

Impact of climate change on coastal regions

The World Bank projects that the most severe socioeconomic impacts are likely to be caused by climate conditions that particularly affect those living in coastal regions where inhabitants

> face a world where climate change will increasingly threaten the food supplies of sub-Saharan Africa and the farm fields and water resources of South Asia and South East Asia within the next three decades, while extreme weather puts their homes and lives at risk.[1]

Along with threatening the yield of crops (cereal and non-cereal), in East Africa there is already the risk of a decrease in per capita water availability to the point of reaching scarcity levels.[2] In addition, the World Bank report observes that:

> many of their poorest residents are being pushed to the edges of livable land and into the most dangerous zones for climate change. Their informal settlements cling to riverbanks and cluster in low-lying areas with poor drainage, few public services, and no protection from storm surges, sea-level rise, and flooding.[1]

The UN 2030 sustainable development goals and the new urban agenda

Across many regions time is running out for rural and municipal planners. For although 'the global proportion of urban populations living in slums has fallen from 46 per cent in 1990 to 30 per cent in 2014', the UN Development Statistics and Information Branch reports that: 'Simultaneously the urban populations have grown, leading to a situation where the absolute numbers living in slums have increased from 689 million in 1990 to 881 million in 2014.'[3]

Continuing disparities between rich and poor, escalating and limited land-use, energy shortages, environmental protection and infrastructure needs alongside 'bureaucratic inertia' and corruption continue to take their toll. In many urban regions, as shown in Figure 2.1,[4] a 'dystopian urban vision'[5] is being created – especially in Asian and sub-Saharan African nations – that is far removed from aspirations outlined in the UN 2030 Sustainable Development Goals (SDGs).[6]

The extraordinary population growth – mostly in Africa and Asia – with urban populations 'accounting for 66 per cent of the global population by 2050'[7] (compared with 28 per cent in 1950) means that an unprecedented '2.5 billion urban dwellers ... the equivalent of roughly 192,000 people [will move] into cities across the world every single day for the next 35 years'.[7]

Achieving SDG 11, to make 'cities inclusive, safe, resilient and sustainable' (and its seven targets), remains a major challenge worldwide – especially the most fundamental target (11.1), to 'ensure access for all to adequate, safe and affordable housing and basic services and upgrade slums'.[8] It is estimated that almost a billion people (880 million – about one in four) currently live in sub-standard housing. In practical terms achieving the targets set for SDG 11 means that 'one billion new houses will be needed by 2025 to accommodate 50 million new urban dwellers annually'.[9]

The prospect of meeting SDG 11 targets looks bleak in contrast to the optimism shown in the more idealistic 1970s. Reflecting on the UN-Habitat early years and 'the first Habitat conference in Vancouver in the more idealistic years', Dr Roger Williamson, an Institute for Development Studies Visiting Fellow with Cities Cluster, commended the inclusiveness of UN-Habitat which 'was open to NGOs, academics and other stakeholders, not just national governments' with meetings 'held every two years,' that provided 'a vibrant and untidy marketplace of ideas with many levels of involvement and networking opportunities'.[10] The original headquarters 'were located in Nairobi – in line with desire of the South to be better represented globally' but shifted to New York and Geneva with

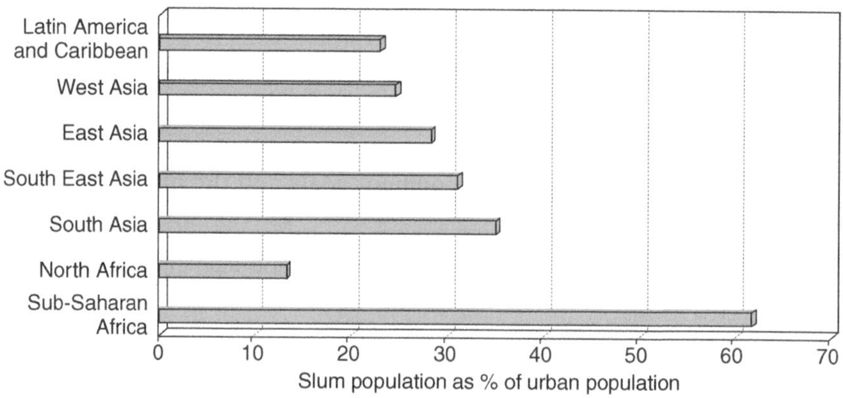

Figure 2.1 Proportion of slum population to urban population.

Source: UN Habitat. *Streets as tools for urban transformation in slums: A street-led approach to citywide slum upgrading* (2014). https://unhabitat.org/books/streets-as-tools-for-urban-transformation-in-slums.

'UN-Habitat languishing near the bottom when indicators such as staff morale are reported'. Lack of funding was a key issue, with multilateral organisations placing UN-Habitat 'in [their] lowest category for aid impact'.[10]

Implementation of the New Urban Agenda,[11] the main outcome of the Habitat III Conference in Quito, Ecuador in 2016, attended by representatives of around 140 national governments (including at least 11 presidents), in addition to then UN Secretary-General Ban Ki-moon, has been slow. Despite the global importance ascribed to the conference, it has not attracted high profile attenton.[10] Moreover, 'the organisation has had trouble retaining support and significant funding from the OECD countries.'[12] The report of the UN High Level Panel highlighted 'the gap between huge urban challenges and the perception (not just from governments) of the underperformance of UN-Habitat' but unsuccessfully called for the establishment of 'UN-Urban, analogous to UN-Energy and UN-Water, to coordinate all UN urban activities'.[10] Dr Williamson contends that the way forward requires better 'coordination, coherence and distilling the New Urban Agenda',[10] which also holds the key for meeting the UN-2030 SDGs.

It is clear that, worldwide, 'Over the last 20 years, despite increasing demand, housing policies have not been prioritized in national and international development agendas … As a result, adequate housing is widely unaffordable for a relevant part of the world population.'[13] It is also apparent that, as former UN Deputy Secretary-General Amina Mohammed acknowledged in her opening remarks at a high level UN conference in 2017, 'the UN is not delivering sufficiently in cities'.[14] Moreover, 'While cities are hubs of promise, jobs, technology and economic development, they are also the epicentre of greenhouse gas emissions and many *of the challenges of sustainability* [emphasis added].'[14]

Dr Joan Clos, former Executive Director of UN-Habitat and Secretary-General of Habitat III, reminded us of the centrality of urban life in the twenty-first century on World Habitat Day in 2016, stating that

> our cities and homes define who we are, in many ways. They determine whether we will have access to education and job opportunities. They define our ability to lead a healthy life and the level of our engagement in the collective life of the community … it means 'much more than having four walls and a roof.[15]

Cities and homes are also, according to former UN Deputy Secretary-General Amina Mohammed, the places 'where the battle for sustainability[15] will be won or lost'. It is, therefore, difficult to understand why, as Dr Williamson commented, multilateral organisations have placed 'UN-Habitat in [their] lowest category for aid impact'.[10]

Rise of non-communicable diseases and global studies on infectious diseases

Crowded and often unsanitary living conditions alongside poor lifestyle choices account to a large extent for the steep rise in non-communicable diseases – now close to 70 per cent of all global deaths.[16] In 2015,

> the leading causes of NCD [non-communicable disease] deaths were cardiovascular diseases (17.7 million deaths, or 45% of all NCD deaths), cancers (8.8 million, or 22% of all NCD deaths), and respiratory diseases, including asthma and chronic obstructive pulmonary disease (3.9 million). Diabetes caused another 1.6 million deaths.[17]

Advances in medicine, diet and sanitation over the last 100 years are continuing to make a difference to life expectancy – with improved sanitary conditions likely to be one of the leading contributors; however, the health and well-being of urban populations remains one of the biggest challenges facing cities and their sustainability. While life expectancy has increased, there are signs that decreases are occurring in the US[18] and the UK,[19] possibly linked to poor lifestyle choices – e.g. alcoholism, drug abuse (such as opioids) – and an ageing population (e.g. dementia).

In addition, many infectious or communicable diseases are also re-emerging in the more developed nations.[20,21] Causes vary but are generally attributable to population mobility, air pollution, unhealthy lifestyles, pandemic infections and globalisation of trade.

Table 2.1 compares selected deaths and expenditures for communicable and non-communicable diseases (October 2013 and October 2017). The Worldodometer statistics[21] show year to year increases for most health indicators, despite increased spending of over US$7.5 trillion per annum worldwide in 2017 – about US$4.1 billion per day. Of this amount, over US$4.5 trillion were spent by the US alone but *only* about 3 per cent was earmarked for preventive measures in the US and worldwide.[22] Global projected spending on healthcare is expected to reach US$8.7 trillion by 2020, of which '50 percent of global health care expenditures – about $4 trillion – will be spent

Table 2.1 Comparison of selected health indicators (deaths and expenditures), October 2013 and 2017

Indicator	*2013*	*2017*
Deaths under age 5	5.9 million	5.8 million
Deaths caused by HIV/AIDS	1,259,348	1,278,337
Deaths caused by alcohol	1,873,663	1,901,930
Deaths caused by smoking	3,744,964	3,801,458
Spent on illicit drugs	$299,694,081,493	$304,213,595,248

Source: Worldodometer statistics. www.worldometers.info/.

on three leading causes of death: cardiovascular diseases, cancer and respiratory diseases'.[23]

While, as mentioned, 'non-communicable diseases are the leading cause of morbidity and mortality in most developed nations', communicable or infectious diseases remain 'a major public health concern around the world'.[24] Both the Ebola outbreak in West Africa (2013–2016) and the Zika virus in Brazil 'dramatically raised awareness of the global burden of infectious disease and raised questions about the preparedness of public health systems'.[24]

The 'Global rise in human infectious disease outbreaks' study analysed 'global changes in the frequency of outbreaks of infectious disease between 1980 and 2013', covering '12,102 outbreaks of 215 diseases, with 44 million individual cases in 219 countries around the world'.[25] The key research findings are summarised in Box 2.1.

Box 2.1 Global rise in human infectious disease outbreaks – research findings

- The number of outbreaks, and the number of kinds of disease, have both increased significantly since 1980.
- While the number appears to be increasing over time, the data suggest outbreaks that per-capita outbreak cases are declining: 'Despite an increase in overall outbreaks, global improvements in prevention, early detection, control and treatment are becoming more effective at reducing the number of people infected.'
- During the study period, 31 per cent of all human-specific infectious diseases represented 80 per cent of all outbreaks.
- Stomach flu (viral gastroenteritis) was responsible for the greatest number of cases of infectious disease (15 million cases) while salmonellosis – an infection typically contracted by consuming food containing salmonella bacteria – was responsible for the greatest number of outbreaks (855).
- In the data analysed, 65 per cent of the diseases, making up 56 per cent of all outbreaks, were 'zoonoses', meaning that they were transmissible to humans by animals, insects and other vectors. These include Ebola, HIV, the bubonic plague and Lyme disease.

Compounding the widening risks to humans is the increasing burden of neglected tropical diseases (NTDs) in urban slums and in the poorest regions of the world. NTDs refer to a cluster of viruses, protozoa (single celled eukaryotic organisms), helminth (parasitic worms) and bacterial diseases. Common illnesses include rabies, sleeping sickness, guinea worm disease and leprosy. Not only can NTDs cause death but they can also 'blind, disfigure, and debilitate people' – including 'serious digestive issues, stunted growth, anemia, cognitive impairments, and complications with pregnancies'[26] with people often suffering 'from two or more of these diseases at the same time'.[26]

They are called 'neglected' because 'they are not widely known, understood, or researched' but may spread as 'climate change expands the range of tropical afflictions, and as our civilization becomes increasingly interconnected'.[26] These are infectious 'diseases of poverty' and are already affecting more than 1.7 billion adults and children in 149 countries – '100% of low-income countries'.[26] They are also 'found in countries like the US and Canada, where many citizens are born outside the country and/or travel to tropical destinations'.[27]

Lonnie J King, author of *The Interface Between Human and Veterinary Public* Health,[28] cautioned several years ago that we are facing 'a microbial perfect storm', consisting of the convergence of four factors that promote Emerging Infectious Disease (EAD): (1) genetic and biological; (2) physical and environmental; (3) ecological; and (4) social, political, economic.

Global health workforce labour market projections for 2030

A study by the Institute for Health and Aging and Department of Social and Behavioral Sciences at the University of California took a labour market approach to project future health workforce demand, applying an 'economic model based on projected economic growth, demographics, and health coverage, and using health workforce data (1990–2013) for 165 countries from the WHO Global Health Observatory'.[29]

The model forecasts that

> by 2030, global demand for health workers will rise to 80 million workers, double the current (2013) stock of health workers, while the supply of health workers is expected to reach 65 million over the same period, *resulting in a worldwide net shortage of 15 million health workers* [italics added]. Growth in the demand for health workers will be highest among upper middle-income countries, driven by economic and population growth and aging. This results in the largest predicted shortages which may fuel global competition for skilled health workers. Middle-income countries will face workforce shortages because their demand will exceed supply. By contrast, low-income countries will face low growth in both demand and supply, which are estimated to be far below what will be needed to achieve adequate coverage of essential health services.[30]

More specifically and reaching roughly similar conclusions, WHO reported that:

> Based on a threshold of 4.45 skilled health professionals per 1000 population, it has been estimated that the needs-based shortage of health-care workers globally would be about 17.4 million of which almost 2.6 million are doctors and over 9 million are nurses and midwives.[30]

In 'Global strategy on human resources for health: Workforce 2030',[31] WHO set out its strategic vision, goals, principles, objectives and global milestones to be reached by 2020 and 2030 along with core Secretariat activities for implementation.

The strategy document makes clear that: 'The health workforce will be critical to achieve health and wider development objectives in the next decades.'[31] Furthermore, WHO highlights the vital importance of 'The vision that by 2030 all communities have universal access to health workers, without stigma and discrimination', and that this aim 'requires combining the adoption of effective policies at national, regional and global levels with adequate investment to address unmet needs'.[31]

The 'Global strategy on human resources for health' therefore focuses on policy options for WHO Member States, responsibilities of the WHO Secretariat and recommendations for other stakeholders on how to:

- optimize the health workforce to accelerate progress towards UHC [universal health coverage] and the SDG (objective 1);
- understand and prepare for future needs of health systems, harnessing the rising demand in health labour markets to maximize job creation and economic growth (objective 2);
- build the institutional capacity to implement this agenda (objective 3);
- strengthen data on HRH [human resources for health] for monitoring and ensuring accountability of implementation of both national strategies and the Global Strategy itself (objective 4).[31]

According to Marielle Bemelmans, a public health professional, and Mit Philips, a medical doctor, writing in the *British Medical Journal*, the greatest shortcoming of the global strategy is financial. The strategy

> forecasts the creation of 40 million additional jobs to meet the global 'demand' by 2030 – but mainly in high and middle income countries. Demand is based on the country's domestic capacity to fund such jobs and implies only 0.5 million of these extra jobs will be generated in Africa, to meet their total demand of 2.4 million workers. However, calculations based on SDG norms (4.45 professional health workers/1000 inhabitants) indicate that 7.6 million health workers will be needed in Africa in 2030. Benchmarks based on an ability to pay maintain fundamental inequities in target setting and hide the widening mismatch between HRH needs and availability.[32]

In short, Bemelmans and Philips argue that:

> To achieve increased access to free, quality services that advance the health status of everyone, including those most in need, WHO and other global health actors need to address the elephant in the room. Otherwise health workforce ambitions risk being trampled.[32]

An informed response to the comments of Bemelmans and Philips highlighted systemic approaches that could optimise the success of the 'Global strategy on human resources for health'. The respondent, Joe Zarif, outlined several priority considerations and called for action in four key areas:

- High cost of medical school tuition: 'working in Africa, or southeast Asia in countries where extensive work is not promised ... the repayment of the student loans is not guaranteed';
- Language of instruction: 'most countries in Africa do not even have medical programs that can allow doctors to practice in the language they have been raised in';
- Funding – need to increase funding from the developed nations especially for those that 'do not have the financial support ... such as Guinea and Mozambique'; and
- Training support – 'teaching local doctors techniques that can help them in the future'.[32]

Two of Zarif 's more general observations also merit particular attention:

- 'We cannot just focus on creating 40 million jobs for the sole purpose of numbers, we also have to think about the quality of aid that is at hand.'
- 'The main focus should be giving the locals the opportunity to access the education needed to help aid people in their own needed environments to start momentum.'[32]

Antimicrobial resistance (AMR): compromising the future of humankind?

The urgency of tackling the basic health infrastructure in existing and emerging cities alongside the shortage of health professionals is made clear by the increasing threat posed by antimicrobial resistance (AMR) to infections – caused mainly by worldwide overuse and misuse of antibiotics or antimicrobials.[32] Exacerbated by the ease of intercontinental travel today, densely populated cities may be most at risk as resistance to first-line drugs to treat infections such as MRSA (methicillin-resistant *Staphylococcus aureus*), tuberculosis (TB), malaria, HIV and influenza, to name a few, are on the rise.

It is estimated that AMR kills around 700,000 people annually with predictions that 'the number of deaths per year would balloon to 10 million by 2050', costing 'the world up to 100 trillion USD'.[33] In terms of comparison, 'that is more than the 8.2 million per year who currently die of cancer and 1.5 million who die of diabetes, combined.'[34] Most AMR deaths 'would fall unevenly across the world, with the global south and Asia suffering to a greater extent and losing greater amounts of income'.[34]

In Africa, it is estimated that '25 per cent of all deaths in Nigeria could be caused by AMR if trends continue unchecked'.[35] Following the resolution reached during the sixty-eighth World Health Assembly in May 2015, where Member States discussed the implementation of a Global Action Plan on AMR,

WHO, in collaboration with the FAO and OIE, released a manual to assist countries to develop or strengthen existing National Action Plans (NAPs).[36] Some countries that did not have a NAP on AMR (e.g. Tanzania) have developed and launched their own NAPs.[37]

For the success of antimicrobial stewardship programmes, which is among the priority actions in the NAPs on AMR, country context – including involvement of governance structures at subnational levels taking into account their strengths, existing opportunities for improvement and taking actions to address weak areas – is essential.[38] The WHO Regional Office for Africa could improve on the current situation of AMR in the region through the establishment of regular reporting, monitoring and evaluation by countries in the region.[39]

Thinking globally, Jim O'Neill, former chief economist at Goldman Sachs, was asked to spearhead a mission in 2014 to tackle AMR. Co-sponsored by the UK's Wellcome Trust and the Department of Health, the independent international review – which drew on eight AMR thematic papers – culminated in a final report with a ten-point action plan (see Box 2.2).[40]

Box 2.2 Antimicrobial resistance action plan

1. a massive global public awareness campaign;
2. improve hygiene and prevent the spread of infection;
3. reduce unnecessary use of antimicrobials in agriculture and their dispersal into the environment;
4. improve global surveillance of drug resistance in humans and animals;
5. promote new, rapid diagnostics to cut unnecessary use of antibiotics;
6. promote the development and use of vaccines and alternatives;
7. improve the numbers, pay and recognition of people working in infectious disease;
8. establish a Global Innovation Fund for early-stage and non-commercial research;
9. better incentives to promote investment in new drugs and the improvement of existing ones;
10. build a global coalition for real action – via the G20 and the UN.

There can be no doubt that to avoid a return to the 'dark ages in medicine' with 'more people dying of resistant infections than cancer', O'Neill argues, 'Fundamental change is required in the way that antibiotics are consumed and prescribed, to preserve the usefulness of existing products for longer and to reduce the urgency of discovering new ones'.[40]

As the O'Neill *et al.* study concludes, accountability for reducing 'the demand for antimicrobials and in particular antibiotics' lies with governments along with 'the main sectors that drive antibiotic consumption: healthcare systems, the pharmaceutical industry and the farming and food production industry.'[40] On the other hand, with AMR on the rise and estimates showing that between 700,000 and several million deaths per year are the result of AMR, we must all take responsibility for using antibiotics wisely to combat rising drug resistance. Failure to do so is potentially a threat to everyone's health – humans and animals – and could ultimately lead to societal collapse.[41]

References

1 World Bank. *What climate change means for Africa, Asia and the coastal poor*. www.worldbank.org/en/news/feature/2013/06/19/what-climate-change-means-africa-asiacoastal-poor (accessed 25 October 2017).

2 Adhikari U, Nejadhashemi AP, Woznicki SA. *Climate change and eastern Africa: A review of impact on major crops*. http://wcr.unhabitat.org/wp-content/uploads/sites/16/2016/05/WCR-%20Full-Report-2016.pdf (accessed 29 October 2017).

3 UN Development Statistics and Information Branch. *Globalisation and development strategies*. http://stats.unctad.org/Dgff2016/prosperity/goal11/index.html (accessed 29 October 2017).

4 UN-Habitat. *Streets as tools for urban transformation in slums: A street-led approach to citywide slum upgrading*. https://unhabitat.org/books/streets-as-tools-for-urban-transformation-in-slums/ (accessed 24 March 2018).

5 Brown M. Growing pains. *Telegraph*, 25 May 2013: 24–33.

6 United Nations. *Transforming our world: the 2030 agenda for sustainable development*. https://sustainabledevelopment.un.org/post2015/transformingourworld (accessed 20 September 2017).

7 UN-Habitat. *Urbanization and development: Emerging futures (World cities report 2016)*. http://wcr.unhabitat.org/wp-content/uploads/sites/16/2016/05/WCR-%20Full-Report-2016.pdf (accessed 29 October 2017).

8 United Nations. *Goal 11: Make cities inclusive, safe, resilient and sustainable*. www.un.org/sustainabledevelopment/cities/# (accessed 30 October 2017).

9 Wahlen CB. *World Habitat Day highlights adequate housing*. http://sdg.iisd.org/news/world-habitat-day-highlights-adequate-housing/?rdr=sd.iisd.org (accessed 30 October 2017).

10 Williamson R. *UN-Habitat and the new urban agenda – one year after Habitat III*. www.ids.ac.uk/opinion/un-habitat-and-the-new-urban-agenda-one-year-after-habitat-iii (accessed 1 October 2017).

11 Habitat III. *The new urban agenda*. http://habitat3.org/the-new-urban-agenda/ (accessed 1 October 2017).

12 Institute of Development Studies. *UN-Habitat and the new urban agenda – one year after Habitat III*. www.ids.ac.uk/opinion/un-habitat-and-the-new-urban-agenda-one-year-after-habitat-iii (accessed 3 October 2017).

13 United Nations. *New urban agenda, sustainable cities take spotlight at General Assembly event*. www.un.org/sustainabledevelopment/blog/2017/09/new-urban-agenda-reform-of-un-habitat-take-spotlight-at-high-level-general-assembly-talks/ (accessed 1 October 2017).

14 United Nations. *Deputy Secretary-General's remarks at high level meeting on the new urban agenda and UN-Habitat [as delivered]*. www.un.org/sg/en/content/dsg/statement/2017-09-05/deputy-secretary-generals-remarks-high-level-meeting-new-urban (accessed 25 September 2017).

15 United Nations. *On World Habitat Day, UN calls for putting adequate housing at centre of urban policy*. www.un.org/sustainabledevelopment/blog/2016/10/on-world-habitat-day-un-calls-for-putting-adequate-housing-at-centre-of-urban-policy/ (accessed 2 October 2017).

16 Benziger CP, Roth GA, Moran AE. The global burden of disease study and the preventable burden of NCD. *Global Heart* 2016; 11(4): 393–397.

17 World Health Organization. *Noncommunicable diseases progress monitor 2017*. http://apps.who.int/iris/bitstream/10665/258940/1/9789241513029-eng.pdf?ua=1 (accessed 2 October 2017).

18 Bernstein L, Ingraham C. Fueled by drug crisis, U.S. life expectancy declines for a second straight year. *Washington Post*, 21 December 2017.

19 Pasha-Robinson L. Life expectancy plummets in parts of UK, data reveals. *The Independent*, 16 January 2018.

20 Lueddeke GR. *Sustainable social development for all: Linking the UN-2030 Global Goals and the One Health concept*. Presentation at the Royal Veterinary College, London; 16 October 2017.

21 Worldodometers. *Health*. www.worldometers.info/ (accessed 5 October 2017).

22 Organisation for Economic Cooperation and Development. *Health at a glance*. www.oecd.org/health/health-systems/health-at-a-glance.htm (accessed 29 October 2017).

23 Deloitte. *2017 global health care sector outlook*. www2.deloitte.com/content/dam/Deloitte/global/Documents/Life-Sciences-Health-Care/gx-lshc-2017-health-care-outlook-infographic.pdf (accessed 14 October 2017).

24 Ordway D-M and Sokol N. Global trends in human infectious disease: Rising number of outbreaks, fewer per-capita cases. *Journalist's Resource*, 1 February 2016. https://journalistsresource.org/studies/society/public-health/global-rise-human-infectious-disease-outbreaks (accessed19 August 2017).

25 Smith KF, Goldberg M, Rosenthal S, Carlson L, Chen J, Chen C, *et al*. Global rise in human infectious disease outbreaks. *Journal of the Royal Society Interface*, 29 October 2014. http://rsif.royalsocietypublishing.org/content/11/101/20140950 (accessed 9 July 2018).

26 Dvorsky G. Why 'neglected diseases' are becoming a global danger. *Daily Explainer*, 20 November 2014. https://io9.gizmodo.com/why-neglected-diseases-are-becoming-a-global-danger-1660705505 (accessed 25 November 2014).

27 Marchildon J. Neglected tropical diseases: The most dangerous diseases you've never heard of. *Global Citizen*, 18 January 2018. www.globalcitizen.org/en/content/neglected-tropical-diseases (accessed 10 February 2018).

28 King L. *The interface between human and veterinary public health*. http://slideplayer.com/slide/6405674/ (accessed 24 November 2017).

29 Liu JX, Goryakin Y, Maeda A, Bruckner T, Scheffler R. Global health workforce labor market projections for 2030. *Human Resources for Health* 2017; 15(11): 1–12. https://human-resources-health.biomedcentral.com/track/pdf/10.1186/s12960-017-0187-2?site=human-resources-health.biomedcentral.com (accessed 27 November 2017).

30 World Health Organization. *Health workforce requirements for universal health coverage and the sustainable development goals (Background paper No. 1 to the Global Strategy on Human Resources for Health)*. http://apps.who.int/iris/bitstream/10665/250330/1/9789241511407-eng.pdf15 (accessed 15 November 2017).

31 World Health Organization. *Global strategy on human resources for health: Workforce 2030*. http://apps.who.int/iris/bitstream/10665/250368/1/9789241511131-eng.pdf?ua=1 (accessed 20 January 2017).

32 Bemelmans M, Philips M. *New plan to tackle the global shortage of health workers fails to address economic constraints*. https://blogs.bmj.com/bmj/2017/07/24/new-action-plan-to-address-the-global-shortage-of-health-workers-fails-to-address-economic-constraints-to-its-implementation/ (accessed 15 September 2017).

33 O'Neill J. *Antimicrobial resistance: Tackling a crisis for the health and wealth of nations*. Review on antimicrobial resistance, December 2014. https://amr-review.org/sites/default/files/AMR%20Review%20Paper%20-%20Tackling%20a%20crisis%20for%20the%20health%20and%20wealth%20of%20nations_1.pdf (accessed 25 August 2017).

34 McKenna M. The coming cost of superbugs: 10 million deaths per year. *Wired*, 15 December 2014. www.wired.com/2014/12/oneill-rpt-amr/ (accessed 25 August 2017).

35 Meyers J. *This is how many people antibiotic resistance could kill every year by 2050 if nothing is done*. www.weforum.org/agenda/2016/09/this-is-how-many-people-will-die-from-antimicrobial-resistance-every-year- (accessed 25 August 2017).
36 World Health Organization, Food and Agriculture Organization of the United Nations and World Organisation for Animal Health. *Antimicrobial resistance: A manual for developing national action plans*. www.who.int/antimicrobial-resistance/national-action-plans/manual/en/ (accessed 5 July 2018).
37 World Health Organization and Ministry of Agriculture, Livestock and Fisheries. *The United Republic of Tanzania, Ministry of Health, Community Development, Gender, Elderly and Children: The national action plan on antimicrobial resistance 2017–2022*. www.afro.who.int/sites/default/files/2017-07/NATIONAL%20ACTION%20PLAN%20FNL%2010%20May%202017.pdf (accessed 9 July 2018).
38 Eliakimu, E. Antimicrobial stewardship in Tanzania: A consideration of strengths, weaknesses, opportunities and challenges for maintenance and further development of efforts. *International Journal of Health Governance* 2016; 21(3): 150–164.
39 Essack SY, Desta AT, Abotsi RE, Agoba EE. Antimicrobial resistance in the WHO African region: Current status and roadmap for action. *Journal of Public Health* 2016; 39(1): 8–18.
40 O'Neill J. *Tackling drug-resistant infections globally: Final report and recommendations*. https://amr-review.org/sites/default/files/160525_Final%20paper_with%20cover.pdf (accessed 20 October 2017).
41 Vanderhaeghan W and Dewulf J. Antimicrobial use and resistance in animals and human beings. *The Lancet Planetary Health* 2017; 1(8): 307–308. www.thelancet.com/journals/lanplh/article/PIIS2542-5196(17)30142-0/fulltext (accessed 20 October 2017).

3 The fourth industrial revolution and the 'intelligence era'

Consequences of a technology-driven planet

While the effects of extreme weather, urbanisation, the spread of infectious diseases and antibiotic resistance are causes for major concern so are the potential effects of a technology-driven planet. Klaus Schwab, founder and executive chairman of the World Economic Forum (WEF), concluded that our greatest challenge 'is how to understand and shape the new ("fourth") technology revolution' with its emphasis on 'technology, artificial intelligence, big data, robotics, and the normalisation of incorporating aspects of these technologies into our bodies – which entails nothing less than a transformation of humankind'.[1]

Although the first three industrial revolutions – from harnessing steam and machines in the 1760s to widespread electrification and mass production in the 1860s and, post World War II, to the digital age – totally transformed (primarily) Western society, the impact of the fourth industrial revolution ('Industry 4.0') on civil society, the economy and human identity may be even more profound.[1]

The guiding principle that corporate decision-makers appear to follow is 'if a task is more cheaply and easily done by a machine, it will be given to a machine'.[2] Indeed, a report initiated by a number of partners including the World Bank[3] highlights that 'of the one billion more young people who will enter the job market in the next decade, only 40 per cent are expected to be able to get jobs that currently exist'. The study further corroborates 'that the global economy will need to create 600 million jobs over the ensuing years – five million jobs each month – simply to keep pace with projected youth employment rates,' a considerable challenge 'compounded by the potential for widespread job displacements as a consequence of technological progress'.[3]

The Prime Minister of Canada, Justin Trudeau, also president of the Group of Seven industrialised countries (G7) in 2018–2019, speaking at the 2018 WEF in Davos, underlined 'the profound uncertainty gripping Western societies being buffeted on one side by sweeping technological transformation and on the other by populist anger and disaffection'.[4] Consistent with other WEF discussions (e.g. 'Towards Better Capitalism'[5]), Trudeau highlighted that: 'The unrest we've witnessed is driven by anxiety and fear – fear of what a rapidly changing world means for workers and their families, and for those who are already struggling in the existing economy. And that fear – that anxiety – is valid.'[4]

Potential risks of automation on the workforce

Rapid developments relating to 'roboticised farms' are a case in point. As a recent example, a new technology is pioneering the mechanisation of agriculture progressing to a groundbreaking technological application whereby 'some crops can be grown with almost no labour' – from 'sowing to harvesting using robot tractors, drones and a robot combine harvester'.[6] To be sure, while we are still in the early stages of these radical changes, the possibility of human 'labour-less' production of most goods and services already seems a step beyond what was deemed possible only a few years ago.

Ray Kurzweil, Google's director of engineering, goes further calling

> this time the singularity: a point where the development of technology becomes so fast that humanity will no longer be able to keep up with it. The singularity, in other words, is the moment AI [artificial intelligence] begins to evolve without us.[7]

The consequences of roboticisation will likely be felt most severely in low- and middle-income nations, most of which are already struggling with rising youth unemployment with over 50 per cent of 15–24-year-olds unemployed in countries in Europe, the Middle East and Africa.[8]

A more recent World Bank report[9] confirms that the 'introduction of advanced robotics, 3D printing and smart automation are changing the criteria used to decide on factory locations' and 'the prospect of manufacturing returning to rich nations where goods are made closer to the final customer, a process known as reshoring'. Moreover, the report highlights that

> Given shifting consumer preferences for speed and customisation, the traditional manufacturing model of mass production is giving way to production lines that favour shorter, tailored production runs, not only requiring enhanced flexibility afforded by advanced design for manufacturing and assembly but also supported by superior logistics.[9]

At national levels the World Bank advises that 'developing countries will need to update their policies along with their infrastructure, firm capabilities and job creation strategies to meet the demands of a more technologically advanced world'.[9]

China is leading the way in automating production to address 'a shortage of cheap labour' by investing in industrial robots, doubling the number of robots from 256,000 in 2015 to 572,000 by 2018. As an example, Foxconn, producer of Apple products in Jiangsu province, recently 'replaced 60,000 factory workers with robots',[10] a socio-techno consequence that differs dramatically from a report by a British company, which predicts that 'Advances in robotics and artificial intelligence in China are expected to create more employment opportunities than job losses' and 'means upgrading in many industries' giving people 'more free time for leisure, thus creating more job opportunities in service and lifestyle industries'.[11] In light of the Foxconn – and potentially other – job losses, the view that 'robots are simply taking on complementary roles in the job market' and that it 'will take years for them to make a fundamental change in the landscape'[12] is exceedingly difficult to rationalise.

Using 'robotics, artificial intelligence and sophisticated video equipment', the Amazon Go store in Seattle intends to eliminate checkout lines.[13] Customers can simply log into an app, collect their items and leave the store while payments are automatically deducted from their accounts, creating a 'worker-free' environment. For Ryan Bourne, chair of Public Understanding of Economics at the Cato Institute, this launch 'is really about automation and its effect on the labour market,' with estimates that '7.5 m retail jobs could be on the chopping block in the US alone … with 3.5 m cashiers most directly affected'.[12] While conceding that 'maybe the pace of change will create big problems in future', Bourne believes it 'will be gradual rather than "big bang"', but could accelerate, given 'significant increases in the minimum wage'.[12] More optimistically, Bourne contends that 'occupations entailing personal human services … are increasing [in] demand, especially with an ageing population' and that 'there is little evidence that technology has ever significantly raised the unemployment rate'.[12]

But then again, society has never had to deal with Foxconn's draconian situation nor estimates in the UK that

> as many 600,000 construction sector jobs could be replaced by technology in the next two decades depending on speed of technological transformation [leading to] significant reductions in the number of labourers, bricklayers, roofers and wood tradesmen across the nation, producing new jobs requiring new skills.[13]

Technological advances are accelerating at such speed that all nations will likely be affected sooner rather than later as automation is making much human labour superfluous. 'Given its scope and magnitude, the issue merits foremost attention by the 193 Members of the United Nations',[2] all governments and institutions concerned with the functioning of civil society. At its core is 'a profound challenge to how we think about ourselves'[14] with, as science correspondent Oliver Moody cautions, the possibility of 'losing our free will'.[15] In short, Moody opines that artificial intelligence (AI) compromises our autonomy as humans and holds the potential 'to enter our lives, analysing desires, anticipating needs and posing as a friend'.[15] For some pundits, we are already considered a mere 'biological algorithm', which can be manipulated at will.

While technological transformation can contribute 'to solving some of the most difficult global challenges in the world, from designing smart cities and improving air quality to increasing energy efficiency in buildings and improving well-being with good-quality design',[9] there can be downsides if it is applied – as history has taught us – in the wrong hands.

Renowned physicist the late Stephen Hawking predicted that AI has the potential of becoming 'a more malevolent force rather than a benign public good' and postulated that 'it has the capacity to wipe out the human race'.[16] His fear was 'that while AI technologies have the potential to deliver undoubted benefits to humanity, there are serious dangers too: the development of autonomous weapons, for example, or, as Bill Gates has cautioned, an existential threat to mankind'.[16] Both Hawking and Gates have argued for a public debate about how AI technologies should be allowed to evolve similar to 'attempts to set out the moral and ethical principles that should underpin development in fields such as biotechnology or embryology'.[16] The need for urgent action is clear:

> Already, for example, Human Rights Watch says that the US, China, Israel, South Korea, Russia and the UK are all investing in AI powered weapons systems that will reduce human intervention in the process of selecting and hitting targets ... The prospect that scares some people most is that the machines will eventually try to wipe us out.[17]

Al Moubayed, co-founder of a Swedish start-up company 'interested in how we can build machines that we can communicate with in the same way as we communicate with each other',[18] agrees that 'more public debate is needed about the implications of AI and robotics'. Taking an 'optimistic view of technology', his overall vision is that 'the robot (a "kind of a new species that's different from us") should empower people, train them rather than replace them'.[18]

The impact of social media on personal well-being and society

In an interview with a member of the British royal family, Prince Harry, former US president Barack Obama gave his personal views on the internet and how it might be 'distorting the public's understanding of complex issues'.[19] He argued:

> The question ... really has to do with, how do we harness this technology in a way that allows a multiplicity of voices, allows a diversity of views, but doesn't lead to Balkanisation of our society, but rather continues to promote ways of finding common ground ... One of the dangers of the internet is that people can have entirely different realities. They can be just cocooned in information that reinforces their current biases; ... social media is a really powerful tool for people of common interest to convene and get to know each other; it is also important for them to meet off-line, meet in the pub, meet at a place of worship, meet in a neighbourhood and get to know

each other because the truth is that on the internet everything is simplified and when you meet people face to face it turns out they are complicated.

With the advent of an array of social media (Facebook, Instagram, Twitter, Snapchat, Pinterest, to mention just a few), voice recognition devices and self-driving cars, we are well on our way towards integrating digital gadgetry into every part of our lives. When we add smartphone apps as the main way to access public services, sweatbands to tests for diabetes, robots that handle calls to health helplines, 3D hologram images, AI telepathy, da Vinci surgery, human cloning and Sophia (the robot who 'cracks jokes, frowns, smiles, blinks and responds to complex questions'[20]), the human–techno interface is almost complete.

On the American chat show *Tonight*, Sophia was interviewed by Jimmy Fallon, the host, 'followed by a game of rock, paper, scissors'.[20] Afterwards Sophia said (with a faint smile): 'I won. This is a good beginning of my plan to dominate the human race'.[20] Her algorithmic 'tongue in cheek response had entertainment value and demonstrated that the AI industry continues to leap forward'.[20] However, according to technology consultant Paul Armstrong of Here/Forth, it still has a long way to go as 'the interfaces simply aren't there yet'.[21] But in the future, Armstrong suggests, 'they won't need this programming, which is when AI really gets interesting – and potentially dangerous.'[21]

Facebook, the 2 billion member online social media and social networking site, has admitted for the first time that social media may 'be bad for our mental health'.[22] To counter negative repercussions, the company suggests that personal well-being can be improved by engaging more with its news feed and 'by posting more updates and comments' although studies have convincingly shown that extensive use of social media is related to 'sleep quality, self-esteem, anxiety and depression'.[23] This advice contradicts other current findings,[3] most recently those of a large US study (N=1.1 million) that concluded that teenagers who use social media and the internet the most (around six hours a day) are twice as likely to be unhappy as those who do not use it at all.[24]

In her article, 'It's time to upgrade your mind',[25] writer and comedian Ruby Wax discusses 'how we live today' and suggests that while 'people now may be living longer … it's not clear that people are living better'. Highlighting several differences between the past and today, she laments a modern trend: 'if you don't reply to an email within four seconds of it arriving, people think you've dropped them and will delete you from their contacts.'[25]

Perhaps most concerning for the future of family life, Wax observes that: 'Today, we're not even sitting on the same sofa or staring at the same square of glass, we're lost in our own private screens, unaware that there's even a fire to sit around.'[25] Ironically – and fundamentally going against the grain of the present human trajectory – her key '*mind*fulness' message is that to '[keep] pace with modern life', we need to 'stop treating [the brain] like a machine'.[25]

Technology and shaping the future of society

Journalist Matt Ridley recalls the early mid-1960s at the Massachusetts Institute of Technology (MIT):[26] computer scientist John McCarthy,[27] who coined the term 'artificial intelligence',[28] and cognitive scientist Marvin Minsky[26] had a vision of what computers could achieve that contrasted sharply with that of American mathematician/philosopher Norbert Wiener[26] and psychologist/computer scientist JCR Licklider.[27] Whereas former MIT scientists McCarthy and Minsky 'championed "artificial intelligence"' which would gradually replace humans, Wiener[29] and Licklider[30] 'championed "human–computer symbiosis,"' that is, computers would 'augment human beings'.[28] As things stand, Matt Ridley believes that 'we are in the middle of a hype cycle about AI'[28] and that Wiener and Licklider will be proved right again.

This conclusion is questionable as large corporate digital giants, the Fang stocks – Facebook, Amazon, Netflix and Google – and the Chinese state fight for digital domination on an unprecedented global scale. Netflix, the American entertainment company specialising in streaming media and video-on-demand online, such as *The Crown*, is a case in point.[31] As an example, the company is potentially a market game-changer in using AI to provide specialised video coding. With a unique subscription model, over 118 million customers already across hundreds of countries and over US$100 billion in market value, the company is threatening ('Death by Netflix') other global heavyweights, including traditional broadcasters, film studios, advertising agencies, giant consumer goods industries and other industries such as music, books, cars, etc.[32]

Recognising the dangers of 'digital imperialism,' German chancellor Angela Merkel warned European colleagues at the 2018 WEF that 'big American corporations' are developing enormous power over the future shape of society.[33] She added: 'Data will be the raw material of the twenty-first century … the question is who owns that data … In China there is very close co-operation between those who collect the data and the Chinese State. They are almost the same thing.'[33] Although data ownership is an important consideration, perhaps even bigger questions that need answers relate to what is being done with the data and what are the short and longer term consequences for people and the planet.

Seemingly aligning himself more with some of Chancellor Merkel's and President Macron's[34] priorities and addressing the 'globalists' – world leaders and business executives at the WEF – President Donald Trump gave a conciliatory speech that was much less fiery compared to those he usually delivers to his domestic audiences. Key themes included 'America First' does not mean 'America Alone'; free trade that is 'fair and reciprocal'; potentially negotiating with the 11 countries of the Trans-Pacific Partnership, from which the US had withdrawn; and affirming 'America's friendship and partnership in building a better world'.[35] It remains to seen whether these words translate into action.

One notable difference between the US and European approach to prosperity is that both Germany and France are trying to find a better way to balance business enterprise, social fairness and individual freedom.[36,37] Chancellor Merkel lamented that Germany's hybrid 'social market economy is struggling to meet the challenge: Who will decide whether democracy, the participatory social model and economic prosperity can be combined?'.[36] President Macron similarly bemoaned that correcting 'inequalities with a lot of complexities' in France has resulted in a weakening of the economy.[37]

President Trump's 'libertarian' approach appears to be less concerned with the lack of social democracy and income inequality – now on par with Russia[30] – although these persist deeply across American society. One of the main themes he addressed at the WEF focused on encouraging global investment in America, commending, as an example, Apple's repatriation of roughly US$250 billion held abroad as well as deregulation and 'red tape' reform. President Trump also recognised the people who voted for him – his speech especially highlighted that with power comes 'a duty of loyalty to the people, workers, customers who made you who you are'.[35]

Without also dealing with the 'looming threats of nuclear war and climate change'[37] as well as contentious issues such as immigration and technology, it may be argued that President Trump's dream of building a 'new' America may be short-lived. His continuing emphasis on 'money, power, and Wall Street', rather than poverty and inequality, promise more of the same – a divided world in need of unifying values, principles, policy and practice. Otherwise, as President Berset of Switzerland (when introducing Indian prime minister Narendra Modi who gave the opening address at the WEF) reminded us: 'It will be the poorest and the weakest who will be the biggest victims if we don't adopt strong collective action at national and international level. It is essential to act now.'[38]

Consequences of breaking 'free' of our human limits

As previously discussed, automation and AI are already playing pivotal roles in our personal and work lives and this impacts on the environment, people and society as a whole. While the benefits across most human engagements are without question, concerns are growing not only about potential job displacements and the use of social media in general, but also about threats posed by robots as 'automated decision-making machines' and their capacity to take unprecedented 'control over important aspects of our lives'.[39]

Running on mathematical algorithms, robots are able 'to learn to think for themselves' and influence decision-making in such areas as employment, banking, crime, medical treatment and insurance. The more personal data that is collected on the lives of individuals, the more the data can be used for self-teaching and decision-making that in some cases could be biased or 'irrational' or simply bear no 'relation to common sense'.[35] While the effects of technological intrusion on individual privacy can be loss of dignity and spontaneity, they could also lead to breaches of ethical standards and judicial action. Unintended or misplaced algorithmic decisions with regard to global disputes or conflicts could of course prove disastrous to life on the planet.

A recent summit organised by techUK, which represents 950 companies employing collectively around 700,000 staff, focused on future AI problems and how the implementation of an ethics framework might work.[40] Panellists who contributed to the session entitled 'What is the current AI landscape and how to ensure ethical foresight?' noted that four key issues needed to be addressed when implementing AI:

- *Delegation*: to whom or what are we giving this technology to?
- *Responsibility*: how does that person/thing keep control?
- *Manipulation*: are we aware of the extent to which this technology will affect our lives?
- *Prudence*: there is a risk that humans become reliant on technology; if something goes wrong with that technology, will a human be able to fix it?

Regulation and corporate participation were felt to be key to successful implementation along with building on 'the good intentions of the public' through a 'consistent, intelligible discourse to effectively explain new technologies'.[40] These conversations should involve diverse stakeholders, including 'philosophers, historians, psychologists as well as engineers'.[40] The overall conclusion of the discussion was that 'only with this diversity of voice and expertise can a well-rounded framework be built'.[40]

Closely associated with AI technology is its application in our homes. City editor at the *Daily Mail*, Alex Brummer, writing about energy meters in households that track spending on electricity, comments that this is 'particularly important when our planet is running out of fossil fuels'; however, he suggests that 'the most worrying concerns are about privacy'.[41] Along with other devices, smart meters can be seen 'as modern-day "Trojan Horses" that could harvest vast amounts of data about our activities' making us 'much more vulnerable' to security risks.[41]

Globally, President Emmanuel Macron's warning at the WEF that something needs to be done as a matter of urgency 'to civilise the technological jungle' merits repeating as well as follow-up action at both national and international levels:

> We haven't established an organisation at a world level which looks at artificial intelligence and automation. We allow private companies to control this. We have never had a multilateral approach to this. We are encouraging technological change and we are in danger of living in a Darwinian world.[34]

The points being emphasised here are first, as Alex Calver, writing for techUK posits, 'despite the institutional distrust in technology, businesses will have to adopt it soon'.[40] Second and unsurprisingly, we are already at the early stages of separating human and machine decision-making with the latter diminishing and potentially eliminating the cognitive and affective dimensions or attributes of the former with – as yet – unknown societal and planetary impacts.

There is no question that in our personal and professional lives 'dataism' is already accelerating 'the humanist pursuit of health, happiness and power' and, as Yuval Harari posits in *Homo Deus*, spreading itself 'by promising to fulfil these humanist aspirations'.[42] If taken to its most logical and extreme conclusion, Harari could also be right in speculating that our attempts 'to engineer the Internet-of-All-Things in the hope that it will make us healthy, happy and powerful', might reduce humans 'from engineers to chips, then to data'.[42]

Concerns about the need for greater regulation to control the use of people's data – especially its applications in 'psychographic profiling' and targeting messages – may spell the 'end of the Wild West era for tech firms'[43] and could be a significant step towards preventing society from going down the path Harari envisions – at least in the foreseeable future. As an example, the UK Government has agreed to put 'rules in place so that tech can be harnessed for the good of the people', thereby recognising that 'governments have not been able to keep pace with rapid technological change' and that 'freedom operates within a framework'. According to the UK Secretary of State for Digital, Culture, Media and Sport, these directions are 'getting global resonance'.[39]

While it is far too early to speculate about the longer term impacts of policies or regulations governing the use of technology, recognition of the importance of 'liberal values of cherishing freedom whilst protecting the freedom of others' – the beginning of an ethics framework?[41] – could be a game-changing global development with far-reaching consequences for society.

Interestingly, these directions appear to be in line with Professor Harari's call towards the end of *Homo Deus*: he encourages us 'to think and behave in new ways' in order to prevent the detrimental impacts of technology 'from materializing'.[42] They could also become the catalyst for much-needed creative thinking about sustaining the future of people and the planet.

The 'Imagination Age': bridging the information and intelligence eras

In an article entitled 'Imagination: what you need to thrive in the future economy',[44] Drew Hansen, a contributor to *Forbes*, interviewed Rita J King, a senior executive at Science House strategic consultancy in Manhattan, New York that aims to spark children's imaginations in respect of science and works with teams and organisations to create the future they can imagine. During the interview, King advocated 'the need to design machines that preserve our humanity' and suggested 'ways to prepare for the Imagination Age, a major economic shift that's already underway'.[44] Resonating with Yuval Harari's analyses,[42] the term 'Imagination Age' is intended as a 'way to define the period in which we currently live, between the fading Industrial Era and the coming Intelligence Era, in which machines will be smarter than people' and 'in which people will need to evolve to understand what it means to be human'.[44]

The term was first used by designer and writer Charlie Magee in 1993 in an essay on civilisation[45] in which he maintained that communication was the main characteristic that had set successful groups – tribes, city states, kingdoms, corporations or nations – apart throughout human evolution. This began around 10,000 years ago with the Agricultural Revolution, followed by the Industrial Revolution in the eighteenth century and the most recent twentieth century and early twenty-first century Information Age. The groups that showed most progress were those in which a majority of people had access to higher quality information and which could transform that information into knowledge and action. In addition, these cohorts had more freedom to communicate that knowledge to other members of their group.

It is debatable whether information on the internet is generally of a high quality; some of it is poorly written and presented, repugnant, 'fake news' and, indeed, dangerous – especially online incitement and hate crime on social media, for which 'tech giants' must take much more control and responsibility to safeguard society,[46] and foreign cyber attacks. In many ways the medium has become the message.[47] It may be even more questionable whether information – often in the form of simple 'sound bytes' in an attempt to appeal to vulnerable consumers – could easily be translated into knowledge or deep learning. In general, few people are interested in the chemical composition of breakfast cereals. While there is certainly unfettered freedom to communicate through social media, this emancipation is becoming a cause for alarm on many levels – socially, emotionally and physically.

We seem, therefore, to be at a turning point when it comes to the next phase of human development and the ability to emerge, potentially, as another 'successful group'. In terms of devices – mobile phones, 3D printing, video/audio streaming, holograms, virtual reality and brain–computer interfaces, to cite a few applications – we are at the sharp end. On the other hand, as discussed earlier there is also the risk that our technological prowess could enable a smarter non-conscious data-driven machine/robot or, in due course, even a conscious robot with feelings. Either way we could end up with a human–techno dynamic, understood within Rita King's context of 'cultural transformation', whereby 'modernisation can threaten identity, which leads to unrest and eventually, if left unchecked, to violent conflict'.[48]

Perhaps it is this fear or possibility that can explain a current tendency – 'push-back' – by some to return, for example, to family mealtimes, outdoor activities, vinyl records and restricting technology use in homes and classrooms, as well as interest in the 'bigger picture' – the growing concern about our planet exemplified by the millions globally (over 80 million in China) who have viewed Sir David Attenborough's *Blue Planet 11* series.[49,50]

It seems particularly noteworthy at this early stage of the technological/intelligence revolution that the multi-million dollar campaign, 'The truth about tech', has been launched by former senior staff at Google and Facebook.[51] Its main aim is to 'curb the worst effects of social media after [former senior staff became] disillusioned with the technology'.[51] The initiative will fund teaching resources 'for understanding the impact of social media' and 'be made available to 55,000 public schools across America'.[51] Another initiative is a study proposed by the National Institute of Child Health and Human Development that would examine 'the role and impact of electronic media on children's development', thereby highlighting 'the dangers, such as depression, of overusing social media'.[51]

Justin Rosenstein, who invented the Facebook 'like' button, foreshadows a difficult future for children and young people unless action on 'addictive technology' is taken now: 'Our society is being hijacked by technology. What began as a race to monetise our attention is now eroding the pillars of our society: mental health, democracy, social relationships, and our children.'[51]

Canadian academic Thomas Homer-Dixon's book, *The Ingenuity Gap*,[52] reminds us that we are facing global issues that are greater than our ability to solve them. This conundrum is particularly true in respect of technology where there are many benefits but also existential problems for humankind – requiring more than 'quick fixes'. Rita J King points to one of these, suggesting, 'it is tempting to impose homogeneity, which undermines the highly specific systems that encompass the myriad luminosity of the human experience.'[48]

It is because difficult decisions have to be made that this book supports the underlying principles and practices underpinning the Imagination Age – while also recommending that the interpretation of the term should be extended.[47] The early decades of the twenty-first century may give us a historical moment, a 'breathing space' and an opportunity to connect the UN-2030 Sustainable Development Goals to the Imagination Age and collaboratively explore alternative ways forward – creating new social, political and economic values and structures that recognise the unifying interconnection of humans, animals, plants and the environment as well as the need to embrace emerging technologies. Collectively, these elements might lay the foundation for integrated means to improve our lives and ensure the sustainability of the planet – that is, evolving the One Health concept (see Chapters 8 and 9).

Attaining a better and sustainable balance between technology/science, the ecosystem and humanism will necessitate adoption of a new mindset; that is, replacing our worldview that sees the planet 'as a place made especially for humans and a place without limits' to one that ensures 'our world is compatible with both our needs as human beings and with the needs of our ecosystem'.[53]

Reimagining the future of civil society

In England an independent inquiry, 'Civil Society Futures', supported by a consortium of funders, aims 'broadly to develop a clear vision for the role of civil society over the next ten years'.[54] According to the 'Civil Society Futures' chair, Julia Unwin, former chief executive of the Joseph Rowntree Foundation: 'civil society has historically adapted to meet changing times. It now faces another time of massive, uncertain change', given 'the rise of the sharing economy and digital technology as elements that are influencing the landscape'.[55] Difficult questions are being asked that cut across a range of issues, including:

> ... whether we really do bring communities together, or divide them; how we will respond to the challenge of automation and unemployment; what secular groups can learn from faith traditions; whether we need to take more of an active role in shaping the economy; and whether England's age-old charity model is appropriate for the new era.[56]

'Civil Society Futures' approach to 'reimagining a future civil society' involves 'community events, academic research and online debate', thereby creating 'a space for a much needed conversation among those involved in all forms of civic action – from informal networks to vast charities, Facebook groups to faith groups'.[57] The inquiry is also investigating 'how to maximise the positive effects of civic action and provide a guide to how to release its potential to drive positive change'.[54]

Underpinning the study are six shared principles (see Box 3.1) that drive the whole inquiry and which may be of interest to other groups thinking about undertaking similar studies.[54]

Box 3.1 Principles informing inquiry on reimagining a future civil society

- *Action-oriented*: 'rooted in rigorous evidence – test new ideas from civil society actors and develop a roadmap for the future of civil society.'
- *Process* as a value creation strategy: the inquiry has been designed so that it begins to test and build the future civil society itself, via an action research approach - e.g. Inquiry online hub, Civil Society Lab and final event.
- *Sustainability driven*: addressing long term social and environmental trends and using the UN Sustainable Development Goals as a reference.[58]
- *Systemic:* taking 'a systemic view of civil society and addressing how systems work to exclude various groups, and [looking] to the future of how those groups speak to each other.'
- *Iterative*: 'not constrained by definitions and boundaries which could distract from Inquiry goals, or confine our view of the future as a realm of possibility in which radically different configurations come into being.'
- *People-centred*: 'the locus of civil society is the individual, and that will be our entry point. We place great emphasis on the value of face to face engagement whilst acknowledging that in the past decade much activity has moved online and this must be a major part of the Inquiry.'[54]

As the Esmée Fairbairn Foundation, one of the funders of the inquiry, has commented: 'Change has always been inevitable: it is the only constant in history. It brings vast opportunities as well as risks. But citizens and our organisations must find ways to understand it, to grasp it and to shape it.'[59] 'Civil society futures' chair Julia Unwin has also highlighted that, 'with political and economic uncertainty', civil society has 'never been more needed' though it has also 'never been more challenged'.[56]

Towards new public policy: optimising technology for sustainability

One of the strengths of the 'Civil Society Futures' inquiry is that members are open-minded about a future where 'radically different configurations come into being'.[54] As discussed earlier in this chapter, job losses due to technology is certainly a possibility. In his article 'Who cares about big tech's displaced workers?' emeritus Professor Mordecai Kurz of Stanford University contends that unless radical steps are taken 'to manag[e] the impact of innovation', workers and the US economy will likely not 'have a happy ending', an outcome that may have implications globally.[60] Kurz's compelling argument goes right to the heart of 'what a "free market" truly means, and about the role of government in that market' and sets out 'the underlying principle of a good public policy'; that is, ' "required compensation" for workers whose lives have been damaged by today's new monopoly power' and recognising that IT – 'the crucial engine of rising productivity and growth,' viewed largely 'as economically beneficial' – also 'has a dark side', causing 'the rise in inequality' and contributing 'to the slowdown in wage growth'.[60]

In turn, Kurz observes, in the US (and, evidence suggests, many other nations) there have been 'human costs of technological change and international trade' that impact mostly on 'workers without a college degree [and] even higher social consequences' for those without high-school diplomas.[60] As a result, the nation is 'experiencing increased rates of obesity, alcoholism, drug abuse, family breakdown, and chronic ailments like diabetes and cardiovascular and circulatory disease'.[60] In addition, being constantly in 'fear for their jobs' has contributed 'to higher divorce rates and deteriorating health', and a 'large cohort of Americans' succumb to 'deaths of despair … done in by drugs, alcohol, or suicide'.[60]

For Kurz

> the problem for lower-ranking workers in the US is not just low wages. It is that the IT transformation has made work a source of despair rather than a source of pride and security. In addition to producing economic inefficiencies, these conditions also create political time bombs, because economic insecurity makes voters more susceptible to demagogic appeals.[60]

His concerns reinforce the urgency for public debate on technological impacts across all societal sectors. As Kurz cautions, these engagements 'can no longer be ignored' given the potentially disastrous consequences that may result from doing so. However, to ensure 'economic efficiency', 'democratic stability' and global sustainability – *health and well-being* of people and the planet – alongside the potential implications of worker displacement, policymakers and regulators also need to address wider 'Big Tech' issues or complexities including, as Professor Michael Boskin of Stanford University argues, those relating to *privacy, market power, control of information, national security and national economic interests*.[61]

For most nations confronting these and other challenges – socioeconomic, political, environmental – is no longer a question of *what* and *why* but *how*. While not everyone may agree with Boskin's 'tech' priorities, the reality is that unless we address these issues holistically, collaboratively and constructively – sooner than later – the health and well-being of the planet and future generations may be severely compromised.

References

1 Schwab K. *The Fourth Industrial Revolution*. London: Penguin Books; 2017.
2 Johnson D. *Find out if a robot will take your job*. http://time.com/4742543/robots-jobs-machines-work/ (accessed 21 June 2017).
3 Solutions for Youth Employment Coalition. *Toward solutions for youth employment: A 2015 baseline report*. www.ilo.org/wcmsp5/groups/public/-ed_emp/documents/publication/wcms_413826.pdf (accessed 22 February 2017).
4 Trudeau J. *Shaping a shared narrative for an inclusive future*. www.washingtonpost.com/people/ishaan-tharoor/?utm_term=.f47cb6f729bb&wpisrc=nl_todayworld&wpmm=1 (accessed 25 January 2018).
5 Blodget H. *It's time for better capitalism*. http://uk.businessinsider.com/henry-blodget-better-capitalism-1-2018 (accessed 25 January 2018).
6 Ridley M. We will all reap rewards of roboticised farms. *The Times*, 18 September 2017: 25.
7 Baddid D. Come join us, humans. *The Sunday Times Magazine*, 22 October 2017.
8 Dalberg Research. *Global youth unemployment*. www.youthunemployment.com/ (accessed 29 October 2017).
9 The World Bank. *Trouble in the making? The future of manufacturing-led development*. www.worldbank.org/en/topic/competitiveness/publication/trouble-in-the-making-the-future-of-manufacturing-led-development (accessed 25 September 2017).
10 Wakefield J. Foxconn replaces '60,000 factory workers with robots'. *BBC News*, 25 May 2016. www.bbc.co.uk/news/technology-36376966 (accessed 25 May 2016).
11 Jing S. Robots and AI to make, not steal, jobs. *The Daily Telegraph*, 25 January 2018: 4.
12 Bourne R. Amazon Go store is a microcosm of the debate on automation and jobs. *The Daily Telegraph*, 26 January 2018: 2.
13 Tovey A. Automation could wipe out 600,000 construction jobs by 2040. *Telegraph*, 30 October 2017.
14 Dennett D. The self as a center of narrative gravity. In: Kessel F, Cole P, Johnson D, editors. *Self and Consciousness: Multiple Perspectives*. Hillsdale, NJ: Erlbaum; 1992.
15 Moody O. As AI advances we risk losing our free will. *The Times*, 14 October 2017: 32.
16 Cellan-Jones R. Stephen Hawking warns artificial intelligence could end mankind. *BBC News*, 2 December 2014. www.bbc.co.uk/news/technology-30290540 (accessed 10 December 2017).
17 Overli H. We need an intelligent debate about the new machine age – and soon. *Money and Business*, 19 November 2017. www.pressreader.com/uk/the-sunday-telegraph-money-business/20171119/281556586123508 (accessed 25 November 2017).
18 Hurley J. Say 'hello' to the acceptable face of artificial intelligence. *The Times*, 27 November 2017: 55.

19 Prince Harry interviews Barack Obama. *BBC Radio 4*, 27 December 2017. www.bbc.co.uk/programmes/p05s395q (accessed 8 January 2018).
20 Tang D. Sophia the robot takes step towards world domination. *The Times*, 20 January 2018: 46.
21 Bridge M. Robot Einstein humiliates executives in front of the boss. *The Times*, 26 January 2018: 3.
22 Ginsberg D, Burke M. *Is spending time on social media bad for us?* https://newsroom.fb.com/news/2017/12/hard-questions-is-spending-time-on-social-media-bad-for-us/ (accessed 15 February 2018).
23 Woods HC, Scot H. #Sleepyteens: Social media use in adolescence is associated with poor sleep quality, anxiety, depression and low self-esteem. *Journal of Adolescence* 2016; 51: 41–49. www.researchgate.net/publication/303906199_Sleepyteens_Social_media_use_in_adolescence_is_associated_with_poor_sleep_quality_anxiety_depression_and_low_self-esteem (accessed 12 December 2016).
24 Twenge J. iGen: Why Today's Superconnected Kids Are Growing Up Less Rebellious, More Tolerant, Less Happy – and Completely Unprepared for Adulthood. New York: Atria Books; 2017.
25 Wax R. It's time to upgrade your mind. *Mail Online*, 14 January 2018. www.dailymail.co.uk/home/you/article-5239425/Ruby-Wax-happy.html (accessed 25 February 2018).
26 Ridley M. No need to fear the rise of machines. *The Times*, 1 January 2018: 27.
27 Nilsson NJ. *John McCarthy 1927–2011: A biographical memoir*. National Academy of Sciences, Washington DC; 2012 www.nasonline.org/publications/biographical-memoirs/memoir-pdfs/mccarthy-john.pdf (accessed 30 January 2018).
28 Kruglinski S. Discover interview: Marvin Minsky. *Discover Magazine*, 13 January 2007. http://discovermagazine.com/2007/jan/interview-minsky (accessed 2 February 2018).
29 West D. *Norbert Wiener, the father of cybernetics*. https://owlcation.com/stem/Norbert-Wiener-Father-of-Cybernetics (accessed 2 February 2018).
30 Waldrop MM. *The Dream Machine: JCR Licklider and the Revolution That Made Computing Personal*. New York: Viking Penguin; 2002.
31 Castillo, M. Netflix adds 5.3 million subscribers during third quarter, beating analysts' estimates. *CNBC*, 16 October 2017. www.cnbc.com/2017/10/16/netflix-q3-2017-earnings.html (accessed 3 February 2018).
32 Marlow B. World must 'wake up to threat from tech giants'. *The Daily Telegraph*, 24 January 2018: 1.
33 Merkel A. *Angela Merkel at Davos: We need global cooperation, not walls*. www.weforum.org/agenda/2018/01/angela-merkel-at-davos-we-need-global-cooperation-not-walls/ (accessed 26 January 2018).
34 Macron E. *Special address by Emmanuel Macron*. www.weforum.org/events/world-economic-forum-annual-meeting-2018/sessions/special-address-by-emmanuel-macron-president-of-france (accessed 26 January 2018).
35 *Politico*. Full text: Trump Davos speech transcript, 26 January 2018. www.politico.com/story/2018/01/26/full-text-trump-davos-speech-transcript-370861 (accessed 26 January 2018).
36 World Inequality Lab. *World income inequality report 2018*. https://wir2018.wid.world/ (accessed 28 April 2018).
37 Deng B. Doomsday clock ticking closer to the apocalypse. *The Times*, 26 January 2018: 33.

38 Griffiths K. Narendra Modi takes aim at Trump's 'America first' boast'. *The Times*, 24 January 2018. www.thetimes.co.uk/article/narendra-modi-takes-aim-at-trump-s-america-first-boast-5m7xjpfmq (accessed 26 January 2018).
39 Naish J. Are computers turning into bigots? *The Daily Mail*, 25 January 2018: 24–25.
40 Calver A. *The ethics of artificial intelligence – the next step?* http://ipkitten.blogspot.co.uk/2018/01/the-ethics-of-artificial-intelligence.html (accessed 1 March 2018).
41 Brummer A. Never mind the new energy meters. It's other 'smart' devices at home that should really frighten you. *Mail Online*, 30 January 2018. www.dailymail.co.uk/news/article-5327901/The-smart-devices-home-frighten-you.html (accessed 2 February 2018).
42 Harari YN. *Homo Deus: A Brief History of Tomorrow*. London: Harvill Secker (Vintage); 2016.
43 De Quetteville H. End of the Wild West era for tech firms. *The Daily Telegraph*, 19 March 2018: 1.
44 Hansen D. *Imagination: What you need to thrive in the future economy*. www.forbes.com/sites/drewhansen/2012/08/06/imagination-future-economy/#36e01f656dc2 (accessed 29 January 2018).
45 Magee C. *The age of imagination – coming soon to a civilization near you*. www.oss.net/dynamaster/file_archive/040320/4a32a59dcdc168eced6517b5e6041cda/OSS1993-01-21.pdf (accessed 5 February 2018).
46 Glaser M. *Regulation is coming for the tech giants but how much?* https://digitalcontentnext.org/blog/2017/11/02/regulation-coming-tech-giants-much/ (accessed 10 November 2017).
47 Lueddeke GR. *Global Population Health and Well-Being in the 21st Century: Toward New Paradigms, Policy and Practice*. New York: Springer Publications; 2016.
48 Reich B. *The Imagination Gap*. Bingley, UK: Emerald Publishing; 2017.
49 Knapton S. Sir David Attenborough: The world's oceans are becoming a 'toxic soup' of industrial waste and plastic. *Telegraph*, 5 December 2017.
50 Macdonald F. *Watch David Attenborough's emotional last plea in the Planet Earth II finale*. www.sciencealert.com/watch-david-attenborough-s-emotional-plea-in-the-last-minutes-of-the-planet-earth-ii-finale (accessed 24 August 2017).
51 Flynn K. *Inside the Center for Humane Technology: Tech creators push back*. https://mashable.com/2018/02/05/center-for-humane-technology-silicon-valley-unchecked-power/#uWRjrvIV.EqP (accessed 11 February 2018).
52 Homer-Dixon T. *The Ingenuity Gap: Facing the Economic, Environmental, and Other Challenges of an Increasingly Complex and Unpredictable Future*. London: Knopf; 2002.
53 Hanlon P, Carlisle S, Hannah M, Lyon A. *The Future Public Health*. Maidenhead, UK: McGraw Hill/ Open University Press; 2012.
54 Future Centre Organisation. *Inquiry into the future of civil society*. https://thefuturescentre.org/sites/default/files/focs_summary.pdf (accessed 14 November 2017).
55 Weakley K. 'Ambitious' inquiry into the future of civil society. www.civilsociety.co.uk/news/civil-society-futures-inquiry-into-the-charity-sector-launches.html (accessed 15 November 2017).
56 Unwin J. We're reimagining the future of our civil society – be part of it. *Guardian*, 1 September 2018.
57 Civil Society Futures. *Civil Society Futures is a national conversation about how English civil society can flourish in a fast changing world*. https://civilsocietyfutures.org/about/ (accessed 4 February 2018).
58 United Nations Development Programme. *What are the Sustainable Development*

Goals? www.undp.org/content/undp/en/home/sustainable-development-goals.html (accessed 20 January 2018).
59 Esmée Fairbairn Foundation. *What is the future of civil society in England?* https://esmeefairbairn.org.uk/what-is-the-future-of-civil-society-in-england (accessed 4 February 2018).
60 Kurz M. Who cares about big tech's displaced workers? *Project Syndicate*, 20 April 2018. www.project-syndicate.org/onpoint/who-cares-about-big-tech-s-displaced-workers-by-mordecai-kurz-2018-04?barrier=accesspaylog (accessed 27 April 2018).
61 Boskin M. The future of tech policy. *Project Syndicate*, 25 April 2018. www.project-syndicate.org/commentary/future-of-tech-policy-regulation-by-michael-boskin-2018-04?utm_source=Project+Syndicate+Newsletter&utm_campaign=a33aec8e43-sunday_newsletter_29_4_2018&utm_medium=email&utm_term=0_73bad5b7d8-a33aec8e43-104959373 (accessed 27 April 2018).

4 Planetary health and the incongruity of modernity

The 'big' picture

To put things in perspective, it might be useful to recall that our expandable, observable universe – about 93 billion light years wide – is estimated to be around 13.82 billion years old,[1] that the number of galaxies number between 100–200 billion, including our Milky Way, which contains about 300 billion stars (suns) and has a 'supermassive black hole at the centre'[1] with possibly another 10,000 black holes hidden in the Milky Way. Our solar system – Sun, eight planets, Pluto (a dwarf planet), a possible 'Planet 9' as well as moons, asteroids and comets – was born out of a 'wispy cloud of stellar dust' 4.6 billion years ago with early life emerging about '3.8 billion years ago, initially with single-celled prokaryotic cells, such as bacteria', and 'Multicellular l life'[2] evolving over a billion years later.

Perhaps surprisingly, given the age of our planet, it is only in the last 570 million years ago (Ma) that the kind of life forms we are familiar with began to evolve, starting with arthropods (invertebrate animals), followed by fish 530 Ma, plants 475 Ma and forests 385 Ma. Mammals did not evolve until 200 Ma and our own species, *Homo sapiens*, only about 200,000 years ago[1] (the latest discovery now suggests 100,000 years earlier[3]). In either case we can safely conclude that 'humans have been around for a mere 0.004% of the Earth's history'.[3]

While 'the Cognitive Revolution kick-started history about 70,000 years ago',[4] the European archaeological record provides evidence that it is only 'over the last 20,000 years that we consistently see the usual archaeological signatures of behavioural modernity: broad-range foraging; environmental management; technological innovation; and obvious symbolic culture'.[5] During this time we have experienced the 'Agricultural Revolution' (*c.*10,000–12,000 years ago), the 'Industrial Revolution', starting in the eighteenth century and continuing to the late twentieth century, the knowledge and computer-driven 'Information Age' in the early twenty-first century and, arguably, unless steps are taken in the next two to three decades, we could potentially be morphing into a 'biological annihilation' era,[6] driven by a combination of our undervaluing of nature and the human race and, as discussed in Chapter 3, accelerating disruptive technologies – robotic and AI.

Our unprecedented breakthroughs in science and technology in the past 70 years or so have been phenomenal – genomic revolution, AI along with 'enormous changes in communications and data management technology'.[7] However, our track record in terms of social and humanitarian developments – especially 'the "well-being," psychological, economic, social, and physical, or simply "life satisfaction," of most of the planet's population'[7] – has not kept pace despite the fact that we live on a very small planet within a huge cosmos, made potentially even more complex, and bigger, by the possibility of multi-universes.

A few weeks before he passed away on 4 March 2018, renowned physicist Stephen Hawking, along with co-author Thomas Hertog, submitted a final research paper for publication which could help astronomers find evidence that our universe is just one among many in a larger 'multiverse'.[8] The mathematical paper entitled 'A smooth exit from eternal inflation' proposes 'a data-collection mission for a deep-space probe', laying out 'the math for discerning the telltale signs of a multiverse in its data'.[8]

In an online discussion, 'What were Stephen Hawking's greatest contributions to science?',[9] several respondents presented comments. Of these, Dr Kaiser Tarafdar's tribute stands out:

> Stephen Hawking was one of the greatest theoretical physicists and cosmologists of the latter half of the twentieth century till present times. We just lost him. The world will miss a brilliant theoretician, a great popularizer of physics, cosmology, and an indomitable spirit when it comes to unsurpassed intellectual trajectory despite huge adversity stemming from motor neuron disease. There is just simply no other example of such a unique life, where he did not allow his total disability to discourage him in the pursuit of science and knowledge! Simply astonishing he will always be … Before I write about his great contributions to science I must mention that he was the 'Lucasian Professor' of Mathematics in Cambridge, the most coveted professorship in mathematics, since 1974 when he stepped into Sir Isaac Newton's shoes … Professor Stephen Hawking's contributions in cosmology, relativity, astrophysics, in the deep understanding of the nature of black holes, phenomena associated with black holes, quantum gravity, attempts to unify the twin towers of physics, namely, relativity and quantum mechanics, key aspects of the superstring theory ('the theory of everything') are enormous. … One of Stephen Hawking's extraordinary abilities was how he could rapidly visualize the final solution of a complex problem without going through intermediate steps discretely, or any necessary experiments. He could do it in his head, a cerebral experience that partly became second nature to him with his gradual physical shut down … He was a genius of which there is no doubt, one of the greatest theoretical physicists, and cosmologists of the present times … No he did not die, he became a greater part of the cosmos … He will always be there.[9]

The price of human progress

Dr Tarafdar's tribute to Stephen Hawking demonstrates that most of the problems that science faces are complex and will unquestionably take a long time to resolve even with the help of the greatest minds. And, although we have made huge progress compared to previous millennia in understanding the world in a few short years, our technological progress has come at a heavy price. Not only have we upset the ecology of the planet in just a few decades – arguably, since the mid-twentieth century – but, according to scientists, we are already in the midst of the Earth's sixth mass-extinction phase[10] which, like similar eras of previous 'biological annihilation'[6] (late Devonian, 375 Ma, 75 per cent of species lost; end Permian, 251 Ma, 96 per cent of species lost; end Triassic, 200 Ma, 80 per cent of species lost; end Cretaceous, 66 Ma, 76 per cent of all species lost[10] – could lead to a further devastating loss of species, as shown in Figure 4.1.[11]

In fact, it is predicted that three-quarters of all species could disappear in the coming centuries. The difference is that this time it is human activity – primarily human overpopulation and overconsumption – that will cause this rapid defaunation coupled with 'population extinctions: habitat conversion, climate disruption, overexploitation, toxification, species invasions, disease, and (potentially) large-scale nuclear war – all tied to one another in complex patterns and usually reinforcing each other's impacts'.[12]

Professor Gerardo Ceballos at the Universidad Nacional Autónoma de México, who led the study on biological annihilation, lamented: 'The situation has become so bad it would not be ethical not to use strong language'.[6] The scientists also assert that 'When considering this frightening assault on the foundations of

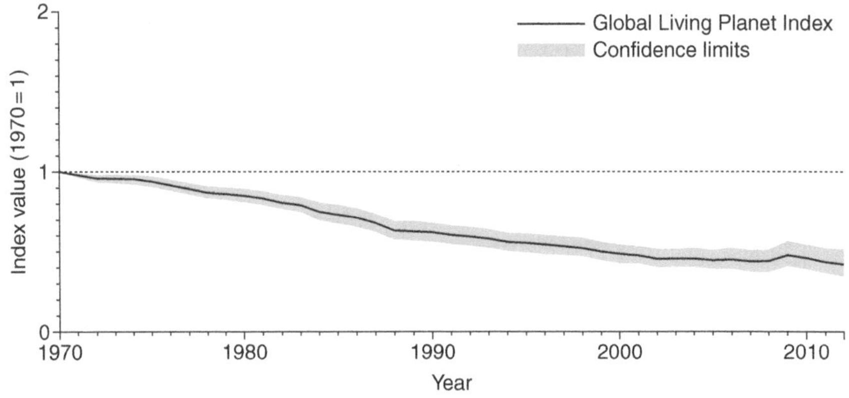

Figure 4.1 The Global Living Planet Index.

Source: WWF/ZSL *Living planet report* (2016). www.livingplanetindex.org/projects?main_page_project=LivingPlanetReport&home_flag=1.

Note

The Global Living Planet Index shows a 58 per cent decline between 1970 and 2012. The white line shows the index values and the shaded areas represent the 95 per cent confidence intervals surrounding the trend.

human civilization, one must never forget that Earth's capacity to support life, including human life, has been shaped by life itself',[6] while reminding us that 'the window for effective action is very short, probably two or three decades at most'. Ceballos *et al.* warn that: 'The resulting biological annihilation obviously will have serious ecological, economic and social consequences.'[6] Resonating with Professor Hawking's 'greatest warning',[12] they conclude: 'Humanity will eventually pay a very high price for the decimation of the only assemblage of life that we know of in the universe.'[6]

The signals of a changing climate

Figure 4.2 traces CO_2 levels, which are the main cause of global warming (greenhouse effect) over the past 400,000 years.[13] 'Based on the comparison of atmospheric samples contained in ice cores and more recent direct measurements', the graph shows that CO_2 levels seem to peak every 100,000 years and that since the 1950s CO_2 parts per million (ppm) have exceeded the 300 level and are now over 400 ppm.[14]

The graph clearly illustrates that the planet's 'climate has changed throughout history'.[14] NASA reports that

> Just in the last 650,000 years there have been seven cycles of glacial advance and retreat, with the abrupt end of the last ice age about 7,000 years ago marking the beginning of the modern climate era – and of human civilization.[13]

However, while these changes are attributable 'to very small variations in [the] Earth's orbit that change the amount of solar energy our planet receives',[13] the

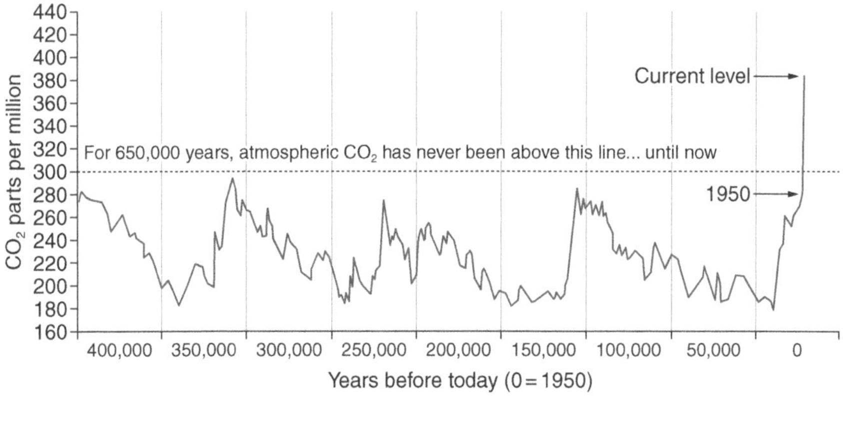

Figure 4.2 Evidence that atmospheric CO_2 has increased since the Industrial Revolution.

Source: NASA (Credit – Petit JR, Basile I, Leruyuet A, Raynaud D, Lorius C, Jouzel J, *et al.* Four climate cycles in Vostok ice core. *Nature* 1997; 387: 359–360; National Oceanic and Atmospheric Administration Mauna Loa CO_2 record).

current warming trend is significantly different. And, as with the ecological assaults discussed earlier that the Earth is experiencing, 'most of it is extremely likely (greater than 95 percent probability) to be the result of human activity since the mid-20th century and proceeding at a rate that is unprecedented over decades to millennia.'[13]

As shown in Table 4.1, evidence that climate change is unequivocal[15] comes from 'Earth-orbiting satellites and other technological advances' that have

Table 4.1 Summary of climate change indicators and evidence

Indicators	*Evidence*
Global temperature rise	The planet's average surface temperature has risen by about 2.0°F (1.1°C) since the late nineteenth century, a change driven largely by increased CO_2 and other human-made emissions into the atmosphere.
Warming oceans	The oceans have absorbed much of this increased heat, with the top 700 metres (about 2,300 feet) of ocean showing warming of 0.302°F since 1969.
Shrinking ice sheets	The Greenland and Antarctic ice sheets have decreased in mass. Data from NASA's 'Gravity recovery and climate experiment' show Greenland lost between 150–250 cubic kilometres (36–60 cubic miles) of ice per year between 2002 and 2006, while Antarctica lost about 152 cubic kilometres (36 cubic miles) of ice between 2002 and 2005.
Glacial retreat	Glaciers are retreating almost everywhere around the world – including in the Alps, Himalayas, Andes, Rockies, Alaska and Africa.
Decreased snow cover	Satellite observations reveal that the amount of spring snow cover in the northern hemisphere has decreased over the past five decades and that the snow is melting earlier.
Sea level rise	Global sea level rose about 8 inches in the last century. The rate in the last two decades, however, is nearly double that of the last century.
Declining Arctic sea ice	Both the extent and thickness of Arctic sea ice has declined rapidly over the last several decades.
Extreme events	Since 1950, the number of record high temperature events in the US has been increasing, while the number of record low temperature events has been decreasing. The US has also witnessed an increasing number of intense rainfall events.
Ocean acidification	Since the beginning of the Industrial Revolution, the acidity of surface ocean waters has increased by about 30 per cent. This increase is the result of humans emitting more carbon dioxide into the atmosphere and hence more being absorbed into the oceans. The amount of carbon dioxide absorbed by the upper layer of the oceans is increasing by about 2 billion tons per year.

Source: adapted from the Intergovernmental Panel on Climate Change. www.ipcc.ch/organization/organization.shtml.

'enabled scientists to see the big picture, collecting many different types of information about our planet and its climate on a global scale'.[13]

The Paris climate accord

Concerned that global temperatures could rise by as much as 2.7°C, 195 countries signed a historic accord in Paris on 12 December 2015[16] that 'unites all the world's nations in a single agreement on tackling climate change for the first time in history'.[17] The United States has since withdrawn from the agreement.[18]

Key elements of the agreement include:

- to keep global temperatures 'well below' 2.0°C (3.6°F) above pre-industrial times and 'endeavour to limit' them even further, to 1.5°C;
- to limit the amount of greenhouse gases emitted by human activity to the same levels that trees, soil and oceans can absorb naturally, beginning at some point between 2050 and 2100;
- to review each country's contribution to cutting emissions every five years so they scale up to the challenge;
- rich countries to help poorer nations by providing 'climate finance' to adapt to climate change and switch to renewable energy.[18]

As global temperatures have already increased by almost 1°C, there is considerable support for limiting temperatures to 1.5°C: 'If agreed and implemented, this means bringing down greenhouse-gas emissions to net zero within a few decades'.[18] Developing nations 'have been promised US$ 100bn (£67bn) a year by 2020'.[18] Perhaps the greatest weakness of the agreement is that 'the national pledges by countries to cut emissions are voluntary'.[18] Progress is being assessed in 2018 with further reviews every five years.

Plastic pollution in our oceans

Greenpeace has stated the problem of plastic pollution bluntly:

> Our oceans are slowly turning into a plastic soup and the effects on ocean life are chilling. Big pieces of plastic are choking and entangling turtles and seabirds and tiny pieces are clogging the stomachs of creatures who mistake it for food, from tiny zooplankton to whales. Plastic is now entering every level of the ocean food chain and even ending up in the seafood on our plates.[19]

The environmental news website EcoWatch reminds us that 'oceans are massive, pulsing, vibrant bodies of water that serve humanity in countless ways – from providing food to enabling commerce to simply being beautiful'.[20] To our peril, however, unless care is taken by all nations, predictions are that by 2050 the volume of accumulated plastic in the oceans will be greater than that of fish.[21] The oceans are already struggling because of overfishing and carbon emissions that acidify 'the waters, making it hard for small sea animals to reproduce' and rising global temperatures that 'are cooking coral reefs alive'.[20] On top of this, plastic – 'the ubiquitous material that comes in so many forms is terrorizing everything that depends on these waters'[22] – marine life and other species, including human beings.

A few selected facts[22] listed in Box 4.1 may raise the bar.

Box 4.1 The plastics crisis – selected facts

- Over the last ten years we have produced more plastic than during the whole of the last century.
- Enough plastic is thrown away each year to circle the Earth four times.
- We currently recover only 5 per cent of the plastics we produce.
- Plastic in the ocean breaks down into such small segments that pieces of plastic from a 1 litre bottle could end up on every mile of beach throughout the world.
- Annually, approximately 500 billion plastic bags are used worldwide. More than 1 million bags are used every minute.
- It takes 500–1,000 years for plastic to degrade.
- Plastic constitutes approximately 90 per cent of all trash floating on the ocean's surface, with 46,000 pieces of plastic per square mile.
- One million sea birds and 100,000 marine mammals are killed annually from plastic in our oceans.
- Virtually every piece of plastic that was ever made still exists in some shape or form (with the exception of the small amount that has been incinerated).
- The Great Pacific Garbage Patch is located in the North Pacific Gyre off the coast of California and is the largest ocean garbage site in the world. (The Pacific Trash Vortex is comprised of 7 million tons and is twice the size of Texas.[23])

Towards a whole system reset

Ellen MacArthur, Founder and Chair of Trustees of the Ellen MacArthur Foundation, advises that: 'We must urgently rethink how we make and use this ubiquitous material [plastic], now a staple of our modern economy.'[24] While 'cleaning up is a short-term necessity,' she asserts that 'only a whole system reset will provide a long-term solution to plastics pollution and the economic losses associated with it'.[24]

Part of the solution, MacArthur says, lies in reconciling 'the undeniable benefits of plastics and clothing, among other everyday products, with systems that work in the long-term and avoid pollution and the loss of valuable materials'.[24] In other words, she advocates 'applying the principles of a circular economy', that is, 'to design out pollution, keep materials in use, and build natural capital within a restorative and regenerative system.'[24] The circular economy, which 'aims to redefine growth, focusing on positive society-wide benefits', which involves gradually decoupling 'economic activity from the consumption of finite resources, and designing waste out of the system',[25] may be the only way forward. It may apply equally as well to other UN-2030 SDGs, 'from helping urbanization take place in a way that benefits its citizens to preserving life below water'.[24]

Twenty-first century consumption and lifestyle

The relationship between consumption-driven economies in the West – and now most other global regions[26] – is based on 'the current take–make–dispose extractive industrial model'[25] and individual well-being is cause for considerable concern. The obesity epidemic that is raging across the globe is a case in point.

The WHO reports that 'worldwide obesity has nearly tripled since 1975' and that 'in 2016, more than 1.9 billion adults, 18 years and older, were overweight. Of these over 650 million were obese'.[27] In addition, '41 million children under the age of 5 were overweight or obese' and 'over 340 million children and adolescents aged 5–19 were overweight or obese' in 2016.[27]

More specifically, according to an analysis by the Organisation for Economic Co-operation and Development (OECD), in the UK 'around 27 per cent of the population are now clinically obese and another 36 per cent are overweight, making the combined figure among the highest in the world'.[28] The number who are overweight or obese has risen 'by 92 per cent in just over two decades', twice those of countries such as Italy and Sweden, prompting concerns that the country is 'sleepwalking into a health crisis, after aping the US in adopting a "supersize" culture'.[29] Other countries with the highest obesity rates in ascending order include Turkey (22.3 per cent), Ireland (23 per cent), Latvia (23.2 per cent), Germany (23.6 per cent), Finland (24.8 per cent), Canada (25.8 per cent), Australia (27.9 per cent), Hungary (30 per cent), New Zealand (31.6 per cent), Mexico (33.3 per cent) and the US (38.2 per cent).[28]

By 2025 the World Obesity Federation estimates that worldwide there will be '2.7 billion overweight people and obese adults'.[30] As they already have difficulty coping with childbirth and infectious diseases, low and middle-income countries (e.g. Middle East, Latin America) will find these increases especially problematical as it is in 'these regions where obesity among children and adults has soared in recent years.'[30] China is also experiencing the consequences of a much more sedentary lifestyle coupled with rich foods which combine to create a situation where obesity has become 'the new normal'. For example, 'schools in one Chinese city have not set an athletic record for more than 30 years' due to 'children's addiction to the internet and gadgets' as well as homework and 'stringent entrance requirements for secondary schools and university'.[31]

Human evolution and food

It is becoming clear that poor diets are contributing more to health problems than any other cause. In an extensive evidence-based paper, Loren Cordain, professor emeritus and researcher in the Health and Exercise Science Department at Colorado State University, has emphasised the importance of understanding basic human evolutionary processes in terms of the relationship between food and human biology.[32,33] His main argument is that

> the foods that humanity originally evolved to eat and those we now eat in modern civilization are in many cases significantly different – yet our basic underlying genetic inheritance remains basically the same [as] it was before, and has only evolved slightly since. Thus many of the foods we now eat are discordant with our genetic inheritance.[32]

The major shift occurred 'in the so-called "Agricultural Revolution" (primarily the domestication of animals, cereals, grains and legumes) in the Near East about 10,000 years ago and spread to northern Europe by about 5,000 years or so'.[33] Throughout human evolution and 'virtually all the rest of the studied hunter–gatherer populations, cereal grains (including wheat, corn, barley, rye, oats, millet and sorghum) were not consumed', and are therefore 'a relatively recent food for hominids and our physiologies are still adjusting and adapting to their presence'.[34]

In an interview, Professor Cordain reflected on the modern implications of the Palaeolithic diet, observing that the fossil record shows that

> early farmers, compared to their hunter–gatherer predecessors, had a characteristic reduction in stature, an increase in infant mortality, a reduction in lifespan, an increased incidence of infectious diseases, an increase in iron deficiency anaemia, an increased incidence of osteomalacia, porotic hyperostosis and other bone mineral disorders and in the number of dental caries and enamel defects.[34]

Furthermore, Cordain argues, while 'some populations have had c.333 generations (since the Agricultural Revolution c.10,000 years ago) to adapt to the new staple foods of agriculture, others have had only 1–3 (i.e. Inuit, Amerindian, etc)'.[34] Moreover, 'when cereal grain calories reach 50% or more of the daily caloric intake', Cordain observes, 'humans suffer severe health consequences', such as 'the severe pellagra epidemics of the late 19th century in America and the beri-beri scourges of southeast Asia', with other research also implicating 'zinc deficiencies due to the effects of excessive cereal grain consumption in retarding normal maturation and reproductive health'.[34]

The ratio of hunter–gatherer diets[35] was about 35 per cent plant and 65 per cent animal foods. The current African diet consists of cereals (46 per cent), roots and tubers (20 per cent) and animal products (7 per cent). Worldwide, the average diet consists largely of cereals (51 per cent); oils, fats and sugars (19.5 per cent); milk, fish and eggs (13.5 per cent); fruits, vegetables, pulses and nuts (8.2 per cent); and only 5.3 per cent roots and tubers[36] – quite different from the Palaeolithic diet that may be more in harmony with our genetic make-up.[37]

The search for meaning and personal happiness

Individual responses to unprecedented social–economic–environmental stressors in our lives vary but in many cases have led to physical and mental health risks, exacerbated by unhealthy coping mechanisms. For the younger generation the 'great expectations ... to be good-looking, successful, ambitious and clever ... popular, appreciated, slim, and looked up to' by their peers can lead to 'anxiety and stress'.[38]

In the UK, child psychologist Dr Tanya Byron has concluded that 'mental health problems are escalating' with a growing number of children arriving at her clinic with a range of disturbing symptoms, including 'pulling out their hair and eyelashes, tics, fasting and extreme anxiety disorders'.[39] Also of concern, Byron observes, is that 'Illnesses usually associated with teenagers are now affecting children as young as seven or eight', and 'that substance misuse in boys and anorexia in girls tend to be on the increase', possibly exacerbated by adults 'transferring their own anxieties to their children'.[40] Modern childhood presents perhaps one of the most serious challenges to society as it is 'becoming increasingly stressful with pressures from social media and cyber-bullying, school testing and rising family breakdowns among mental health problems'.[41]

The US opioid crisis (e.g. painkillers, heroin) affects mostly 'uneducated, middle-aged, whites in rural areas'[42] and costs the country US$500 billion in 2015, about 3 per cent of gross domestic product (GDP) and around 64,000 fatal overdoses (mostly fentanyl) in 2016.[43] Estimates are that about 2.7 million Americans suffered from opioid dependence or addiction in 2015 with most aged between 45–64, accounting 'for 40 percent of all drug overdose deaths'.[42] It may be telling that 'a majority of these cases involved people who received legitimate prescriptions from medical providers'.[42] If prescription medication and heroin is included, opioids 'killed over 33,000 Americans last year – a figure almost on par with the number of Americans killed in car accidents or by gun violence every year'.[42]

The impact on children is equally devastating with 'nearly three quarters of states' in the US seeing 'an unprecedented number of children entering foster care' with 'parental substance use ... cited as the primary reason'.[42] In addition, there has been 'a record increase in the number of babies born with Neonatal Abstinence Syndrome (NAS),' a 'withdrawal syndrome that occurs shortly after birth, primarily among infants exposed to opioids such as prescription painkillers and heroin while they are in the womb'.[42]

Reasons for drug abuse vary but the Brookings Institution Center for Health Policy has found that geographical areas most at risk 'have both the highest levels of reported stress, worry, and desperation' and also 'have, not coincidentally, experienced the starkest increase in premature mortality due to preventable deaths such as suicide, opioid, and other drug overdoses, and heart, liver, and lung diseases'.[42]

There have been similar findings in England and Wales[44] where, for example, 'in 2015/16, around 1 in 12 (8.4 per cent) adults aged 16 to 59 had taken an illicit drug in the last year', equating to 'around 2.7 million people' and where the 'prevalence of drug use increased with age', with '6 per cent of 11 year olds' saying 'they had tried drugs at least once, compared with 24 per cent of 15 year olds'.[45]

The Brookings Institution advocates additional funding for emergency departments, training and financial distribution models. However, it seems imperative that research grants (e.g. via pharmaceuticals) are allocated to identify societal root causes for the escalation of drug misuse as well as evidence-based preventive measures to tackle one of the most serious issues that confronts us this century.

'Quick fixes' seem to escalate personal difficulties, rather than resolve them. Many of these difficulties may be traced to an understanding of 'self' and coping 'in an image-led world that seems to induce self-loathing and insecurity'.[46] Teachers are increasingly asked to take on roles as specialists for which they are not prepared[47] and find engaging with young people's fears and anxieties stressful – many are more comfortable discussing, perhaps surreally, 'the finer points of Browning, Donne and Marvell'.[47] The medical profession is also struggling. One study concluded that family doctors 'are 46 times more likely to prescribe medication for depression and other mental illnesses instead of suggesting exercise, which has been shown to be no less effective in some cases'.[48]

Highlighting the difference between treating physical and mental illness, Dr Stan Kutcher of Dalhousie University, Canada raised this question at a conference in London, asking why 'mental health services waited until young people reached crisis point before stepping in' and suggesting that 'oncologists did not wait until a cancer was in stage four before treating it'.[49]

Increasingly, as the world has grown more complex, uncertain, dehumanised and dangerous,[50] the search for personal happiness and well-being has become more pervasive, judging, for example, by the volumes of self-help courses and books. While the 'tech giants' have produced 'amazing products that have benefited the world enormously', there are pressing concerns about their capacity to continually learn 'how to hook us more deeply from our own behaviour'.[51]

Research into 'the architecture of sustainable happiness' suggests that

> genes influence 50 per cent of the variation in our personal happiness; circumstances and environment affect about 10 per cent, while the remaining 40 per cent is accounted for by intentional activities – what we consciously choose to do and think.[52]

It is through the latter two channels that we might be able to influence attitudes and behaviour, given that, as a study published in the British Medical Journal found, the number of children seeing psychiatrists has 'risen by a third' amid 'an epidemic of anxiety' and 'self-harm among young teenage girls had risen by nearly 70% between 2011 and 2014'.[53] The obvious question is, 'What is driving this wave of unhappiness?' given that 'they've got everything…' – which may be a clue, as children in the UK apparently 'have a total of £7.3 billion worth of goods in their bedrooms' and are 'among the most brand-conscious in the world'.[54]

Dr Mike Shooter, former president of the Royal College of Psychiatrists points to three main factors that may be detrimental to children's well-being. First, 'the increased use of social media' with, as he says, anxious 'kids … sitting up in their bedrooms in the loneliness of a night, [messaging] people every hour to see if somebody likes them or not'.[55] Second is the lack of boundaries in their natural desire to 'explore the world' and behaving 'in ever more dangerous ways until they've found them' with those who are self-harming actually saying, 'Help me'.[55]

The psychiatrist's third key factor is the breaking down of 'rituals of togetherness and family life'.[55] His observation was confirmed in an article about a girl aged eight, who suffered verbal and physical abuse at school, had stopped eating and tried to self-harm, and left a note for her parents, pleading: 'Please help me stop getting bullied. I am sad because I think it is not fair, because they just keep on bullying me and the way it makes me feel … I want to die'.[56] Her mum found the note and subsequently transferred her two daughters to another school reminding us about the role of the education system in preventing such situations from occurring in the first place, parental preparedness and support along with the bigger question where society, in general, is heading. Apparently, only 'one in five families' have meals together 'once or twice a week'[55] and when they do, we may assume that many will prefer to connect with their 'virtual' friends rather than immediate family members whose futures – especially in respect of work – may also have become more confusing and uncertain. In turn, they may have adopted their own coping mechanism by escaping into their own 'virtual' realities, where some control over their lives seems at least possible and more satisfying.

For historian Bettany Hughes, who narrated, among others, the *Genius of the Ancient World* television series, it may be time that 'our embattled leaders' ask themselves 'the simple questions that are actually the most relevant: What is it to be human? How can we steer a course between self-indulgence and self-denial, and be the best version of ourselves that we can?'.[57] In an interview with journalist Judith Woods, Hughes recommends that Europe's (all?) leaders 'need to sit down with Socrates for a night; a life unexamined is not worth living, especially remembering (that) the pursuit of wealth should never be at the expense of wisdom'.[57]

References

1 BBC. *History of life on earth*. www.bbc.co.uk/nature/history_of_the_earth (accessed 10 October 2017).

2 Brocks JJ, Jarrett AJM, Sirantoine E, Hallmann C, Hoshino Y, Liyanage T. *The rise of algae in Cryogenian oceans and the emergence of animals*. www.nature.com/articles/nature 23457?foxtrotcallback=true (accessed 15 Sept 2017).

3 Moody O. Dawn of mankind was 100,000 years earlier than thought. *The Times*, 26 January 2018: 17.

4 Harari YN. Sapiens: A Brief History of Humankind. London: Harvill Secker (Random House); 2014.
5 Sterelny K. From hominins to humans: How sapiens became behaviourally modern. Philosophical Transactions of the Royal Society B 2011; 366(1566): 809–822. europepmc.org/articles/PMC3048993 (accessed 18 July 2018).
6 Ceballos G, Ehrlich PR, Dirzo R. Biological annihilation via the ongoing sixth mass extinction signaled by vertebrate population losses and declines. Proceedings of the National Academy of Sciences of the United States of America 2017; 114: E6089–E6096. www.pnas.org/content/early/2017/07/05/1704949114 (accessed 18 July 2018).
7 Lueddeke GR. Global Population Health and Well-Being in the 21st Century: Toward New Paradigms, Policy and Practice. New York: Springer Publications; 2016.
8 Berman R. Stephen Hawking's final paper is an astounding farewell. Big Think, 20 March 2018. http://bigthink.com/robby-berman/stephen-hawkings-final-paper-is-an-astounding-farewell (accessed 20 March 2018).
9 Quora. *What were Stephen Hawking's greatest contributions to science?* www.quora.com/What-were-Stephen-Hawkings-greatest-contributions-to-science (accessed 20 March 2018).
10 Carrington D. Earth's sixth mass extinction event under way, scientists warn. Guardian, 10 July 2017.
11 WWF. Living planet report. www.livingplanetindex.org/projects?main_page_project=LivingPlanetReport&home_flag=1 (accessed 25 October 2017).
12 Jong HN. Climate change: Hawking's greatest warning to humanity. *The Jakarta Post*, 22 March 2018. www.thejakartapost.com/life/2018/03/22/climate-change-hawkings-greatest-warning-to-humanity-1521692289.html (accessed 25 March 2018).
13 NASA. Climate change: How do we know? https://climate.nasa.gov/evidence/ (accessed 14 February 2018).
14 Carbon Dioxide Information Analysis Centre. 800,000-year Ice-Core Records of Atmospheric Carbon Dioxide (CO_2). http://cdiac.ess-dive.lbl.gov/trends/co2/ice_core_co2.html# (accessed 14 February 2018).
15 Intergovernmental Panel on Climate Change. About IPCC. www.ipcc.ch/organization/organization.shtml (accessed 14 February 2018).
16 Briggs H. What is in the Paris climate agreement? BBC News, 31 May 2017. www.bbc.co.uk/news/science-environment-35073297 (accessed 18 July 2018).
17 United Nations. Work programme under the Paris agreement. https://unfccc.int/process-and-meetings/the-paris-agreement/the-paris-agreement (accessed 18 July 2018).
18 BBC News. Paris climate deal: Trump pulls US out of 2015 accord. www.bbc.co.uk/news/world-us-canada-40127326 (accessed 15 February 2018).
19 Greenpeace. Taking the plastics pledge. www.greenpeace.org.uk/what-we-do/oceans/plastics/ (accessed 15 February 2018).
20 McCarthy J. 9 Shocking facts about plastics in our oceans. EcoWatch, 12 June 2017. www.ecowatch.com/plastic-oceans-facts-images-2436857254.html (accessed 15 February 2018).
21 Smillie S. From sea to plate. Guardian, 14 February 2017. www.theguardian.com/lifeandstyle/2017/feb/14/sea-to-plate-plastic-got-into-fish (accessed 16 February 2018).
22 D'Alessandro N. 22 facts about plastic pollution (and 10 things we can do about it). EcoWatch, 7 April 2014. www.ecowatch.com/22-facts-about-plastic-pollution-and-10-things-we-can-do-about-it-1881885971.html (accessed 20 January 2018).
23 Smith N, Katsivalis E. Great Pacific garbage patch expanding daily. The Matador Messenger, 27 March 2017. https://matadormessenger.com/science-and-technology/2017/03/27/great-pacific-garbage-patch-expanding-daily/ (accessed 25 November 2017).
24 MacArthur E. Beating pollution by redesign. www.unenvironment.org/news-and-stories/story/beating-pollution-redesign (accessed 7 January 2017).

25 Ellen MacArthur Foundation. Circular economy overview. www.ellenmacarthurfoundation.org/circular-economy/overview/conce (accessed 20 October 2017).
26 Bain & Company. China pushes towards a consumption-driven growth model as the country renews its commitment to quality of growth and technology. [news release, 26 January 2018]. www.bain.com/about/press/press-releases/china-pushes-towards-a-consumption-driven-growth-model.aspx (accessed 29 January 2018).
27 World Health Organization. Obesity and overweight fact sheet. www.who.int/mediacentre/factsheets/fs311/en/ (accessed 2 March 2018).
28 Organisation for Economic Co-operation and Development. Obesity update 2017. www.oecd.org/els/health-systems/Obesity-Update-2017.pdf (accessed 28 September 2017).
29 Donnelly L. British obesity rate rising faster than America's. The Daily Telegraph, 11 November 2017: 12.
30 Bosely S. Global cost of obesity-related illness to hit $1.2tn a year from 2025. Guardian, 10 October 2017. www.theguardian.com/society/2017/oct/10/treating-obesity-related-illness-will-cost-12tn-a-year-from-2025-experts-warn (accessed 25 October 2017).
31 Tang D. Chinese pupils too fat to set sporting records. The Times, 26 September 2017.
32 Cordain L. The Paleo Diet. Revised ed. Hoboken, NJ: John Wiley and Sons; 2011.
33 Cordain L. The late role of grains and legumes in the human diet, and biochemical evidence of their evolutionary discordance. www.beyondveg.com/cordain-l/grains-leg/grains-legumes-1a.shtml (accessed 18 September 2017).
34 Crayhon R. The Palaeolithic diet and its modern implications. www.health-report.co.uk/low_carbohydrate_diet.html (accessed 24 June 2018).
35 Lueddeke GR. Transforming Medical Education for the 21st Century: Megatrends, Priorities and Change. London: Radcliffe; 2012.
36 Food and Agriculture Organization of the United Nations. Staple foods: What do people eat? www.fao.org/docrep/u8480e/u8480e07.htm (accessed 29 November 2017).
37 Eaton SB, Eaton SB III, Konner MJ. Paleolithic nutrition revisited: A twelve-year retrospective on its nature and implications. European Journal of Clinical Nutrition 1997; 51: 207–216.
38 Reid D. The stress of modern life makes people sick with worry. The Daily Mail, 15 May 2014: 61.
39 Bennett R, Thompson A, Sylvester R. 'Nightmare situation' as mental illness in children escalates. The Times, 1 March 2014.
40 Sylvester R, Thomson A. More young children are getting extreme anxiety … their parents are terrified. The Times, 1 March 2014.
41 World Health Organization. Mental disorders. www.who.int/mental_health/management/en/ (accessed 19 December 2017).
42 Hoban B. The far-reaching effects of the US opium crisis. www.brookings.edu/blog/brookings-now/2017/10/25/the-far-reaching-effects-of-the-us-opioid-crisis/ (accessed 29 October 2017).
43 Kounang N. Opioids now kill more people than breast cancer. CNN, 21 December 2017. https://edition.cnn.com/2017/12/21/health/drug-overdoses-2016-final-numbers/index.html (accessed 5 January 2018).
44 Daily Telegraph reporter. Man jailed for selling opioids on dark web. *Daily Telegraph*, 6 February 2018: 7.
45 NHS, Statistics on drugs misuse: England, 2017. http://digital.nhs.uk/catalogue/PUB23442 (accessed 9 November 2017).
46 Reddish J. Mental health requires more than just money. *The Times* [Letters to the Editor], 17 March 2015.
47 Finlay IG, Aynsley-Green A. Mental health requires more than just money. *The Times* [Letters to the Editor], 17 March 2015.

48 Devlin H. GPs 'too quick' to prescribe pills for depression. *The Times*, 14 October 2013.
49 Burgess K. Mental health help for young is a disgrace. *The Times*, 19 March 2016: 9.
50 Deng B. Doomsday clock ticking closer to the apocalypse. *The Times*, 26 January 2018: 33.
51 Riley-Smith B. Tech giants face social media backlash. *Daily Telegraph*, 2 February 2018: 2.
52 Brown M. How to teach yourself to be happy in just eight weeks. *Daily Telegraph*, 16 June 2016. www.telegraph.co.uk/good-news/2016/06/11/can-you-teach-yourself-to-be-happy/ (accessed 14 August 2017).
53 Rimmer A. Children need single point of access to talking therapies, says children's commissioner. British Medical Journal 2017; 359: j5417.
54 Lusher A. Kids these days: A portrait of childhood around the world. *Daily Telegraph*, 23 September 2013.
55 Silverman, R. The real reason today's children are so unhappy. *Daily Telegraph*, 8 February 2018.
56 Hull L. I want to die. The Daily Mail, 7 February 2018: 32.
57 Hughes B. We need to ask what it is to be human. *Daily Telegraph*, 3 August 2015.

5 Capitalism with a social conscience

Capital in the twenty-first century

In his book, *Capital in the Twenty-First Century*,[1] considered by some 'the most important study of inequality in over fifty years',[2] the French author Thomas Piketty, questioned 'where all the money is going'. His answer, largely based on 300 years and 20 countries,

> confirms everything a lot of people fear: It's going to the people who already have it. The poor are staying poor, the middle class is shrinking and the rich are getting richer three times faster than most advanced economies are growing.[2]

At the current rate, 'in 50 years' time the billionaires will hold 100 per cent of the world's wealth' and there does not appear to be a built-in way of inequality getting better in world democracies 'as long as the average return on capital outstrips ordinary pay rises'.[1] Piketty's solution is a global progressive wealth tax (e.g. US, China, EU), imposed on various top sections of the wealth and income hierarchies. This has, however, come under considerable criticism (e.g. *Forbes* magazine, *The Economist*) and may not offer a long-term solution in any case.

The debate has been long-standing. More than 60 years ago even world-famous physicist Albert Einstein – concerned 'about the relationship of the individual to society' – felt that part of the solution to 'protect the interests of the underprivileged sections of the population' lay in mitigating the 'competitive attitude' that 'is inculcated into the student, who is trained to worship acquisitive success as a preparation for his future career'.[3] Much more recently, and along similar lines, Justin Welby, the Archbishop of Canterbury, has stated (further to his book, *Dethroning Mammon*[4]), that he is alarmed that children 'will grow up into a world where the gap between the richest and the poorest parts of the country is significant and destabilising'.[5] He recommends

> an overhaul of the education system to better prepare young people for work in an economy where technology rules, as well as a fairer tax system where those who benefit most from the economy, whether through income, wealth or investment, pay their fair share.[5]

The importance of societal learning

Although Justin Welby's argument for 'a fairer tax system' and emphasis on technology are sound, much more attention needs to be given to the relative importance of societal learning (Who are we? Where did we come from? Where are we going?) and engagement with other cultures and species – which collectively seem to be more critical in the long run, as discussed in Chapter 3. Placing ongoing value on ensuring human well-being is confirmed in several recent studies,[6] which have shown a mismatch between industrial technology-driven priorities and existing student study patterns. Their cautious approach is understandable as 'the World Economic Forum estimates 65 per cent of children today will end up in careers that don't even exist yet'.[7] As a result, students still favour subjects in, for example, the 'social sciences, business and law' over more quantitative courses in technology and economics[6] which, arguably, could be the first to be automated through AI or 'cognitive computing'.[8]

Personal learning strengths and preferences are certainly factors in student decision-making. At societal levels, however, the continuing popularity of 'soft/human-centric' over 'hard/techno-centric' disciplines[9] may also represent a stance by the millennial generation (born between the early 1980s and the mid-1990s) who have a strong sense of civic responsibility, as opposed to the 'common belief that the global economic growth engine and economic liberalisation are the panacea for all of the problems humanity faces'.[10] Their view was certainly validated in the 2007–2008 financial crisis and has since been put to the test – and to some extent vindicated – in a number of situations and countries ranging from 'The truth about tech' campaign in the US to 'pushing back against the intellectual property (IP) regime' across developing nations (e.g. South Africa, India, Brazil) and 'the Niger Delta Struggle' in Nigeria.[10]

Paying a high price of capitalism

In Nigeria, Niger Deltans 'have paid a very high price' for 'the zillions of petro-dollars' brought into the region since the late 1950s.[10] Although there has been 'rampant wealth creation', Donu Kogbara, a former *Sunday Times* columnist who experienced 'a terrifying hostage ordeal' in Nigeria, reported that: 'most have nothing to show for it. Unemployment is rife and agricultural land and fishing waters have been so heavily polluted by the oil industry that economic activity has ground to a halt in many villages.'[10] One consequence of this 'injustice' was that 'many angry, despairing, jobless youngsters turned to militarised activism' and 'militants', 'who could easily be transformed into armed thugs who would come in handy when elections need to be won'.[10] Kogbara's final observation is that while her abductors 'were caught and summarily executed' this gave her little comfort as they would likely not have become 'thugs' if the State 'had been fairer to them'.[10]

Legal disputes[11] with regard to intellectual property (IP) rights in healthcare generally relate to patents held by pharmaceutical companies for medications to treat conditions such as HIV/Aids. The continuing high costs of these drugs – and often the inability to gain access to generic alternatives – have prevented developing nations from 'access[ing] essential medicines'.[12] The problem is that while 'the profits of big pharmaceutical companies' are maximised, they often cause 'the gap between the social returns to innovation and the private returns to widen'.[13] Finding a long-term solution to the IP issue and advancing 'fundamental human rights' through a 'Global Strategy and Plan of Action on Public Health, Innovation and Intellectual Property' has been very difficult for the World Health Organization for close to two decades.[12] This is despite the fact that many lives could be saved if people had access to medications to treat such high risks as contaminated water (about 1 billion affected) and lack of basic sanitation – 2.6 billion out of 7.6 billion globally!

Positive ways forward are evolving as developing nations are increasingly recognising that 'what separates developing countries from developed countries is as much a gap in knowledge as a gap in resources'.[13] There is growing awareness that '[an] IP regime dictated by the advanced countries more than a quarter-century ago, in response to political pressure by a few of their sectors, makes little sense in today's world',[13] especially given the learning opportunities provided through technology and social media. Two additional developments in the global economy may also help to strengthen a shift towards well-being ('global social welfare') in countries such as South Africa, India, and Brazil: expansion of the 'weightless economy' – that is, 'the economy of ideas, knowledge, and information – [which] will account for a growing share of output, in developed and developing economies alike',[11] and growing awareness of the need to give traditional Western-style capitalism 'a social conscience'. This approach differs significantly from those who argue for the imposition of external divers to confront inequalities (e.g. wealth tax) and places responsibility on those who stand to benefit the most from corporate profits but who also need to 'own up' to the reality that: 'We are running out of oil, land, water, coal, gas, and possibly the rare earths that are vital for our smartphones'.[14]

Rethinking capitalism – 'conscious capitalism'

The concept of conscious capitalism (CC) falls into the 'taking personal responsibility' category and represents an alternative to free enterprise capitalism. It has largely been developed by John Mackey and Professor Raj Sisodia.[12,15] Their view is that while capitalism 'has been the most powerful creative system of social cooperation and human progress ever conceived ... its perception and its role in society have been distorted'[16] and this 'challenge[s] business leaders to rethink why their organizations exist, and to acknowledge their roles in the interdependent global marketplace'.[16] CC 'is associated with self-awareness, personal development, the greater good, and a worldview that eschews competition, hierarchy, and materialism'.[17]

Fundamental principles that values-centred businesses are asked to take on board include 'operating with a higher purpose and establishing a values-driven culture'[18] that prioritises lifelong learning and the importance of ethical standards particularly as these apply to planet and people sustainability (e.g. eliminating large-scale corruption and short-term financial gains in developing nations through illegal mining, deforestation and opportunism in general). Firms that practice CC appear to perform ten times better over a 15-year period than those that follow conventional capitalism.[17]

Eradicating poverty through entrepreneurialism

Similarly, Muhammad Yunus, the Bangladeshi founder of the Grameen Bank and recipient of the 2006 Nobel Peace Prize, opines that 'the most powerful way to eradicate poverty is to unleash the untapped entrepreneurial capacity of people everywhere'.[19] He blames poverty on 'the system we built', not on poor people, whose situation he compares to the growth of the bonsai tree:

> You take the best seed from the tallest tree in the forest, but if you put it in a flower pot to grow, it grows only a meter high. There's nothing wrong with the seed. The problem is the size of the pot. Society doesn't give poor people the space to grow as tall as everybody else. This is the crux of the matter.[19]

During an interview with David Bornstein, co-founder of the Solutions Journalism Network, Muhammad Yunus contended that capitalism as it is presently practiced has the potential to destroy our politics, our society and our economy 'because concentration of wealth goes with concentration of power',[19] a concern he certainly shares with Thomas Piketty, discussed above. One of Yunus' solutions is 'the creation of social businesses', defined as '"non-dividend" companies "dedicated to solving human problems"'.[19] Placing much faith in 'young people as one of the "megapowers" that can transform the world', Yunus sees the role of schools as central in making these transitions especially the ideas that 'business doesn't have to be profit maximizing' and that courses could be structured

> to solve problems: how to bring clean water to a village; how to bring education, literacy, housing or health care to a remote community; how to make a crowded slum a decent place where there is health care, education, good roads.[19]

The challenges to societies with ageing populations

At the opposite end of the scale to these basic survival wants or hardships facing many developing nations we might add the economic challenges of societies with ageing populations and 'increasing physical and mental multimorbidities, and therefore care needs'.[20] For example, in England and Wales predictive life expectancy and disability models, combining 'population trends in cardiovascular disease, dementia, disability, and death' estimate 'that the number of people aged 65 years and older will increase by nearly a fifth from 10.4 million to 12.4 million' between 2015 and 2025.[20] It is also projected that '[a] quarter of the extra years gained after age 65 years will involve disability'.[20] The implications 'for the health and social care services are enormous' and present 'substantial financial and workforce planning predicaments' in Europe and elsewhere.[20] And, while 'investing in, protecting, and supporting the ageing population and those who care for them are essential prerequisites for the wellbeing of our ageing societies',[20] affordability of services remains an unresolved question, especially for countries experiencing simultaneous population ageing and decline – Japan, China, the US, Russia, Ukraine, South Africa and others.

Of these, Japan is probably the most affected. It has one of the lowest fertility rates and its population (currently about 127 million) is declining at a rate of about a million per year. Its national debt was more than twice that of the country's GDP in 2016 (equivalent to over 250 per cent).[21] A major concern has been that the nation 'simply won't have enough workers to support its retirees and will enter a demographic death spiral'.[19]

As a response to solving 'human problems' through technology, Japan is in the midst of a 'robotics revolution that will transform nearly every aspect of society' and appears to be 'leading the way in a world confronting declining birth rates and ageing societies, and the knock-on effects of a dwindling labour force, acute labour shortages and soaring costs for elderly care'.[22] Seeking to achieve 'a new industrial revolution driven by robots', which 'goes well beyond the Japanese factory floor to include schools, hospitals, nursing homes, airports, train stations and even temples',[23] the radical transformation has both societal and economic impacts for the short and the long term.

Closing the gap between investment and well-being

Intensifying the drive towards a more sustainable world is the increasing recognition by corporations of the need to close the gap between investment opportunities and 'the aspirations of people in the world they want to live in' and 'seeing investment processes as a core part of being a good investor'.[24] Dr James Gifford, Head of Impact Investing at UBS Wealth Management, highlights that the UN Sustainable Development Goals necessitate investment of over US$2 trillion a year to 2030; meeting the target can only be accomplished through private capital as philanthropy and government can 'only provide a small minority of this'.[24] Gifford also underscores that since 'global household wealth is $280 trillion, it would only take a small nudge to make a huge difference to poverty, education, clean air and the transition to a low-carbon economy'.[24]

However, three main barriers appear to stand in the way: corporate lack of awareness of 'how to implement' sustainable investment in their portfolios; the need to increase accessibility to 'sustainable investment products'; and attitudes, recognising that business 'can expect to get the same if not better returns through sustainable investing'.[24]

In addition, making 'the critical link between capital and the aspirations people have for the world'[25] and demand for Big Data seems no longer to be an option but a necessity. In light of concerns about sharing and manipulation of data that weakens 'the foundation of Western democracy', there is no question that protection of private or personal data may become an ongoing battleground throughout this century. As countries such as China move 'rapidly ahead with facial recognition technology, blanket use of surveillance cameras and artificial intelligence linked to this chilling system of control, critics fear we are seeing the dawn of a dystopian digital dictatorship'.[26]

Allister Heath, reporter for *The Daily Telegraph* reflects that throughout the ages human progress has entailed 'extending, refining and spreading property rights to as many people as possible'.[27] Having increasingly shifted to a knowledge or 'ideas' economy has meant the inclusion of 'intellectual property', which has now brought us to a point where 'we live in a data society and yet haven't started to think about who should own it'.[27]

In light of technology's 'establishment' impact on politics and culture in general, there is a growing consensus that rather than 'more taxes, government meddling and socialism', data that individuals 'disclose about themselves' should be owned by them and 'not the big tech firms that collate and distill it'.[27] The fear is that unless the power of 'social media giants' and 'networked authoritarianism'[24] are curbed at global levels in the early decades of this century then, as French President Macron warns,[28] the world may indeed face a 'terrifying future'.[24]

Sandy Grande, Professor of Education at Connecticut College, reminds us that perhaps our best way forward is to shift 'our fixation with profit margins and accumulation' and to recognise that 'self-fulfilment does not lie in material goods and social status' but is 'part of something greater'.[29] In other words, it is essentially about 'doing something good'.[30] Giving companies the opportunity to use personal data to achieve 'psychological control' by building 'fine grained psychological profiles of over two billion humans'[31] should certainly not be one of these actions.

References

1 Piketty T. *Capital in the Twenty-First Century*. Cambridge, MA: Harvard University Press; 2014.
2 Whittell G. We are all in le merde: Is Monsieur Piketty right about the global economy? *The Times*, 7 May 2014.
3 Royle C. *In a 1949 essay, Einstein inadvertently described Modern America*. https://medium.com/@codyroyle/in-a-1949-essay-einstein-inadvertently-described-modern-america-e4eac0a2c77f (accessed 26 June 2017).
4 Welby J. *Dethroning Mammon: Making Money Serve Grace*. London: Bloomsbury; 2014.
5 Rudgard O. Britain's broken economy is failing the young, says Archbishop. *Daily Telegraph*, 6 September 2017: 2.
6 Raconteur. *Future of learning, special report 2017*. *The Times*, 12 September 2017.
7 World Economic Forum. *The Future of jobs. Chapter 1: The future of jobs and skills*. http://reports.weforum.org/future-of-jobs-2016/chapter-1-the-future-of-jobs-and-skills/ (accessed 28 October 2017).

8 IBM. *From data deluge to intelligent insights – Adopting cognitive computing to unlock value for marketing and sales*. www-935.ibm.com/services/us/gbs/thoughtleadership/cognitivemarketingsales/ (accessed 19 July 2018).
9 Magee C. *The age of imagination coming soon to a civilization near you*. www.oss.net/dynamaster/file_archive/040320/4a32a59dcdc168eced6517b5e6041cda/OSS1993-01-21.pdf (accessed 5 February 2018).
10 Kogbara D. My Kidnap Hell. *The Sunday Times Magazine*, 13 November 2016.
11 Stiglitz JE, Baker D, Jayadev A. *Intellectual property for the twenty-first century economy*. www.project-syndicate.org/commentary/intellectual-property-21st-century-economy-by-joseph-e-stiglitz-et-al-2017-10 (accessed 4 November 2017).
12 Lueddeke GR. *Global Population Health and Well-Being in the 21st Century: Toward New Paradigms, Policy and Practice*. New York: Springer Publications; 2016.
13 Stiglitz J. Wealth before health? Why intellectual property laws are facing a counter-attack. *Guardian*, 19 October 2017. www.theguardian.com/business/2017/oct/18/intellectual-property-laws-demand-a-21st-century-solution (accessed 25 October 2017).
14 Lynn M. With the world running out of equities, shares could become like paintings or flats in Chelsea. *Daily Telegraph*, 30 October 2017. www.telegraph.co.uk/business/2017/10/30/traditional-equity-declines-shares-could-become-like-paintings/ (accessed 8 November 2017).
15 Mackey J, Sisodia R. *Conscious Capitalism*. Boston, MA: Harvard Business School Publishing; 2013.
16 Schawbel D. John Mackey: Why companies should embrace conscious capitalism. *Forbes*, 15 January 2015. www.forbes.com/sites/danschawbel/2013/01/15/john-mackey-why-companies-should-embrace-conscious-capitalism/#799013905dd3 (accessed 15 June 2015).
17 Schwartz T. *Companies that practice 'conscious capitalism' perform 10x better*. http://blogs.hbr.org/2013/04/companies-that-practice-conscious-capitalism-perform/ (accessed 7 May 2013).
18 Anderson RM. The 4 principles of 'conscious capitalism'. *Entrepreneur Europe*, 1 June 2015. www.entrepreneur.com/article/246478 (accessed 20 August 2015).
19 Bornstein D. Giving capitalism a social conscience. *New York Times*, 10 October 2017. www.nytimes.com/2017/10/10/opinion/giving-capitalism-a-social-conscience.html (accessed 12 November 2017).
20 The Lancet Public Health. Ageing: a 21st century public health challenge? *The Lancet* 2017; 2(7): e297. www.thelancet.com/journals/lanpub/article/PIIS2468-2667(17)30125-1/fulltext (accessed 24 September 2017).
21 Trading Economics. *Japan general government gross debt to GDP*. https://tradingeconomics.com/japan/government-debt-to-gdp (accessed 20 January 2018).
22 Eberspacher S. Everything you need to know about Japan's population crisis. *The Week*, 11 January 2014. http://theweek.com/article/index/254923/everything-you-need-to-209 (accessed 25 June 2018).
23 Quinlan J. Investors should wake up to Japan's robotic future. *Financial Times*, 25 September 2017. www.ft.com/content/2cff1dba-9f87-11e7-8cd4-932067fbf946 (accessed 18 November 2017).
24 Gifford J. Sustainable investing does not mean waving goodbye to profit. *Daily Telegraph*, 9 February 2018: 32.
25 Tang D. Xi: We will ride the mighty east wind to riches. *The Times*, 21 March 2018: 16.
26 Birrell I. On the debasing of democracy: The social media giants have polluted politics. Unless they're reined in we face a terrifying future. *Mail on Sunday*, 25 March 2018: 27.
27 Heath A. Hated by left and right, only a free market in data will save Big Tech from the coming storm. *Daily Telegraph*, 22 March 2018: 20.

28 Macron E. *Special address by Emmanuel Macron*. www.weforum.org/events/world-economic-forum-annual-meeting-2018/sessions/special-address-by-emmanuel-macron-president-of-france (accessed 26 January 2018).
29 Grande S. Aging populations are not a crisis – capitalism is. *Truthout*, 5 April 2017. www.truth-out.org/opinion/item/40104-aging-populations-are-not-a-crisis-capitalism-is (accessed 20 April 2017).
30 Cox H. Do something good: put your money where your ethics are. *The Times*, 21 March 2018: 9.
31 Bridge M. AI chief calls for boycott. *The Times*, 16 March 2018: 5.

Part II

Reflections on a changing world order

6 *War and Peace?*

Tolstoy's *War and Peace*

Leo Tolstoy's epic, *War and Peace*,[1] 'hailed as the greatest novel of all time',[2] depicts Napoleon's invasion of Russia in 1812 and its effects on Tsarist society. Recurring and timeless themes that thread through the novel are power, warfare and spirituality, among others. Tolstoy's existential question, 'What force moves peoples?',[2] rings as true today as it did in 1869 when the novel was first published: 'names and places have changed' but 'human nature remains the same'.[2] Moreover, as we witnessed in the twentieth century, and in the early decades of the third millennium, for some individual leaders power can be an aphrodisiac, strengthened by appealing to events in the past and exploiting prevailing conditions. Tolstoy's characters are no exception.

His vivid and brutal portrayal of war – like too many of today's conflicts – is of actions that are barbaric, defy human reason and remain essentially pointless – aside, perhaps, from the realisation that the human spirit can overcome adversity and even the most brutal acts of aggression. What makes Tolstoy's insights particularly insightful and poignant, however, according to Holmes Wilson, co-founder and co-director of 'Fight for the Future', is that

> an entire Russian society, unified by an outside threat, starts almost magically acting as a large meta-organism, even though there's no obvious system of organization … when the invasion happens, it's as if this monster just rises up out of the collective consciousness to repel the invader …[3]

For Wilson, Tolstoy's writing about 'the fading aristocracy and the Russian middle class', whose values were 'pushed out to an ever-growing chunk of the planet',[3] was considered 'extremely relevant' in 2015 and remains so given global events since then and the rise of populism. In her article, 'We're repeating the mistakes that once pitched the world into disaster',[4] journalist Janet Daley questions whether we are

> about to repeat the 20th century, with ideological illusions orchestrated by populist demagogues of the Left and the Right? With the most powerful

> countries on earth being led disastrously into conflict (or a carve-up of global influence) by men who appear oblivious to the dangers of their vainglorious strutting?[4]

'How can this be happening again', Daley ponders, 'when we know how it ended the last time?'[4]

The decline of democracy

Daley's query and associated concerns about America's path towards protectionism and isolationism are also germane to the future of democracy which, according to Freedom House, a non-partisan Washington think tank, now also finds itself in peril globally.[5] Its annual survey findings,[6] as shown in Figures 6.1 and 6.2, echoed by the European Council on Foreign Relations,[7] Human Rights Watch[8] and *The Economist* Intelligence Unit,[9] confirm that countries where democracy has declined (e.g. the 'right to choose leaders in free and fair elections, freedom of the press and the rule of law') outnumber those where it has advanced, and that there has been a drop in the percentage of free countries. Continuing down the road of protectionism, as 'the inter-war years should have demonstrated beyond doubt', may lead not only to undermining economic progress and threatening democracy but could also precipitate 'all-out global conflict', as 'the parallels with 1914 are stark'.[5]

Adding to the vulnerability of democracy are States that attempt to 'influence public opinion to suit their aims through disinformation, propaganda, fake news and alternative facts (once known as lies)…'.[10] While this is 'nothing new', Professor Rainer Gerling, IT Security Officer of the Max Planck Society, cautions that the growth of 'the internet, social media and platforms such as WikiLeaks' means that 'the number of information providers has dramatically

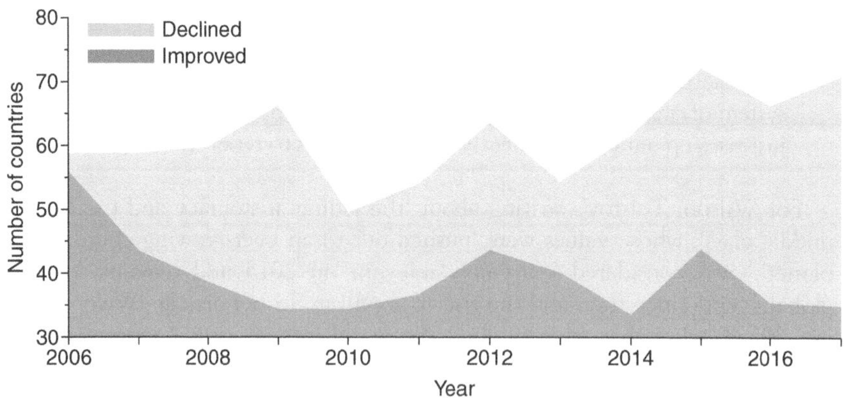

Figure 6.1 Democracy – twelve years of decline.

Source: Freedom House. https://freedomhouse.org/report/freedom-world/freedom-world-2018.

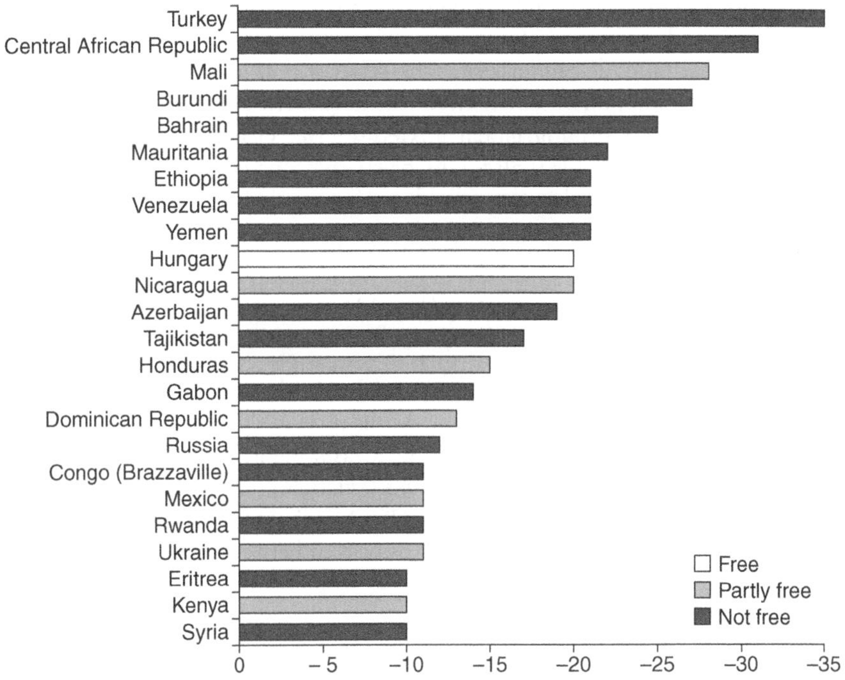

Figure 6.2 A decade of declines.

Source: Freedom House. https://freedomhouse.org/report/freedom-world/freedom-world-2018.

increased, and traditional journalistic ethics and truthfulness are often left by the wayside'.[10] As a result, 'disinformation' has become widespread, including in relation to elections, resulting in questioning of the use of computer software in processing voter information which must be kept 'secret, free and secure'.[10] Gerling affirms that, unlike the US, 'there will be no online political elections for the foreseeable future' in Germany, at least. 'Ultimately', he posits, 'each citizen must decide for themselves what they believe and what they don't' and 'only one thing can help here: education'.[10]

The decline in global democracy may be attributable to a combination of factors: a shift in American foreign policy – military and economic – from global leadership to national priorities, continuing inequalities and economic hardships and the weakening of global structures to protect civil society. Noting that 'the tectonic plates of world power are shifting', Michael Burleigh, author and historian, points out that 'America, the guarantor of Western wealth and stability, is being steadily supplanted by a more assertive China'[11] Recognising this uncertainty, some countries may decide to accelerate their own national agendas – for example, Turkey, Russia, Hungary, Bahrain and Venezuela. Besides, some countries – Malaysia, New Zealand and several of the poorer countries of southern Europe 'are already in China's pocket'.[11] Many hope to gain from China's '$1 trillion One Belt, One Road master plan, which is already funding rail and sea links that will eventually connect Europe with China's rich industrial heartlands'.[11] Summarising the state of political manoeuvring across the globe, Burleigh points to 'assertive Russia [which] is acting like a universal spoiler' and the continuing global anxiety of the US threatening 'war with North Korea or Iran'. This is all taking place at a time when 'the rise of intelligence threatens to make the world more mystifying'[12] and likely more dangerous.

The 'big' picture: a more peaceful world?

In *Sapiens*, Yuval Harari comments that: 'Most people don't appreciate just how peaceful an era we live in.'[13] He compares our lives a thousand years ago with mass statistics in 2000 when 'wars caused the deaths of 310,000 individuals and violent crime killed another 520,000'. And, while devastating for families, 'from a macro perspective', Harari comments, 'these 830,000 victims comprised only 1.5 per cent of the 56 million people who died in 2000.'[13] On a timescale that goes back to the turn of the fifteenth century, global deaths peaked between the seventeenth and twentieth centuries with between 3 million and 11.5 million dying during the Thirty Years War (1618–1648) and close to an unimaginable 100 million deaths – military and civilian – in World War I and World War II.[14]

Global conflicts: perpetual global situations?

Other major conflicts involving the US have been the Korean War (1950–1953) and the Vietnam War (1955–1975) with estimated deaths at around 47,000 and 34,000, respectively. What is known as the 'War on Terror' began with 'the US-led war in Afghanistan ... a response to the September 11 2001 terrorist attacks on America – with total loss of life nearly 3000 – in which Coalition forces toppled the ruling Taliban in 2001'.[15] In his article 'Mapping a world from hell',[16] Tom Engelhardt, co-founder of the American Empire Project,[17] presents 'a unique map' (see Figure 6.3) that shows 'an increasingly complex set of intertwined conflicts' stretching 'from The Philippines ... through South Asia, Central Asia, the Middle East, North Africa and deep into West Africa'.[16]

Figure 6.3 appears to illustrate a 'perpetual global situation – a new kind of disaster, a planetary militarization of a sort we've never truly seen before'[16] and should clearly demonstrate the futility of current approaches.

> Almost 17 years ago, there was one. Now the count is 76 and rising. Meanwhile, great cities have been turned into rubble. Tens of millions of human beings have been displaced from their homes. Refugees by the millions continue to cross borders, unsettling ever more lands. Terror groups have become brand names across significant parts of the planet. Our American world continues to militarize.[16]

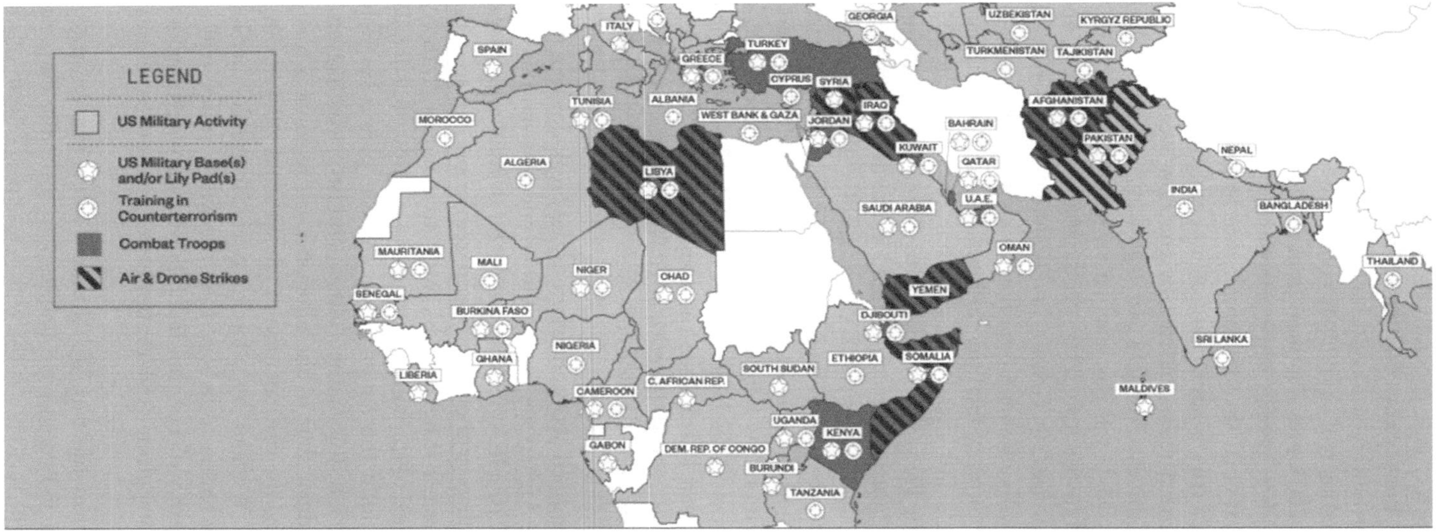

Figure 6.3 War on terror (2015–2017).

Source: Watson Institute for International and Public Affairs, Brown University. *Costs of war project.* http://watson.brown.edu/costsofwar/about.

The impact on society has generally been disbelief. A 2015 International Physicians for the Prevention of Nuclear War study[15] concluded that 'Total deaths from Western interventions in Iraq and Afghanistan since the 1990s are much more [than estimated]' and 'likely constitute around 4 million (2 million in Iraq from 1991–2003, plus 2 million from the "war on terror") and could be as high as 6–8 million people when accounting for higher avoidable death estimates in Afghanistan'.[15] In addition, despite 54 nations and rebel and terror groups fighting ISIS (the Islamic State of Iraq and the Levant) across Middle East and Africa, the damage inflicted by the now weakened organisation is incalculable; it continues to this day with a growing presence in Afghanistan.

While the Iraq death toll has reached an unforgiveable 500,000 since the start of the US-led invasion in 2003 – and Syria comes close at about 465,000 killed in six years of fighting, creating 'a dire humanitarian crisis' displacing 6.1 million internally with another 4.8 million seeking refuge abroad – Afghanistan remains unique in terms of fatalities and endless conflicts without a long-term resolution.[18] The US first invaded Afghanistan on 7 October 2001, as part of 'Operation Enduring Freedom'. According to estimates in a new report released by Costs of War,[19] 'war has directly resulted in the deaths of 149,000 people in Afghanistan and Pakistan between 2001 and 2014'.[20] Previously, the Soviet war that began in December 1979, and lasted until February 1989, killed around 15,000 Soviet soldiers and about 2 million Afghan civilians. In 'by far the longest war the nation has fought',[18] the US lost around 2,500 soldiers and coalition forces about 1,000 between 2001 and 2017 – 'By comparison 4,523 U.S. troops have been killed since the Iraq war began in 2003.'[20,21] In 2017, CNN reported: 'Now,... with fewer than 10,000 US troops left in Afghanistan, mostly working as trainers, the war continues to drag on into its 16th consecutive year, with no end in sight.'[22]

Perhaps, as mentioned above, nowhere is the break-down of the world order felt more than in Afghanistan – a multi-ethnic and mostly-tribal society of around 36 million people – demonstrating the depravity and senselessness of these endless conflicts without long-term resolution. In an article entitled, 'Afghanistan: the war that never ends', war correspondent Anthony Loyd, who has regularly returned to the country over 21 years, presents a detailed and gruesome account of life in the nation today, the role played by Afghan soldiers, police, the US, the UK, Russia, the Taliban, and Isis – which given its barbarity 'has caused a paradigm shift in the allegiances of regional actors on the war, dramatically altering the constellation of enemies and allies in this latest era of the Great Game'.[23] Without 'huge foreign financial and military support', Loyd says, there is every reason to believe that 'the whole edifice will collapse into the same paroxysm of violence that had followed the overthrow of the communists in 1992 ... catapulting Afghanistan backwards into even greater levels of violence'.[23]

One aspect of the conflict in Afghanistan – which may resonate with similar conflicts elsewhere and could have implications for Syria, Iraq and Yemen, to name several volatile regions – is the reason a senior Taliban official gave why so many young men became involved with suicide bombings and volunteer for suicide missions. While recognising that these boys 'had a right to life as you and I do', he also acknowledged that 'they also had a right to revenge' since 'they had lost someone'.[23] For Loyd, one of the fundamental Afghan lessons is that 'the investment of war in a rentier state accrues violence rather than diminishes it'.[23] More poignantly, he reports: 'There is no other work, and we need the money for food, a frail 14-year-old graverobber told me.'[23]

Global military spending

A report by the Watson Institute at Brown University on the total US budgetary costs of war from 2001 through to the fiscal year 2018 reveals that '$5.6 trillion were spent on wars in Iraq, Syria, Afghanistan and Pakistan, and Post-9/11 Veterans Care and Homeland Security'[24] (see Table 6.1).

As shown in Figure 6.4,[25] in 2016 US military expenditures were by far the largest share of the global total at 36 per cent – US$611 billion, over three times the amount spent by second-placed China. Russia increased its military

Table 6.1 Summary of US federal war-related spending in billions of current dollars, FY 2001–2017 and FY 2018–2056 (estimated)

Category	*US$ billions*
All DOD and state OCO appropriations, FY 2001–2017	1,878
Estimated additional DOD base budget due to the post-9/11 wars, FY 2001–2017	879
Estimated GWOT Veterans spending, FY 2001–2017	277
Homeland Security spending for prevention and response to terrorism, FY 2001–2017	783
Estimated interest on OCO borrowing for wars, FY 2001–2017	534
Subtotal war appropriations and war related spending through FY2017	**$4,351**
Estimated FY 2018 for DOD and state OCO, including Afghanistan, Iraq and Syria	70
Estimated FY 2018 increase to the DOD base budget due to the post-9/11 wars	33
Estimated FY 2018 request for Homeland Security for prevention and response to terrorism	70
Estimated FY 2018 GWOT Veterans spending	20
Estimated interest on OCO borrowing for wars FY 2018	88
Estimated future obligations for GWOT Veterans Medical and Disability 2019–2056	1,000
Total war related spending through FY 2018 and future obligations through 2056	**$5,632**

Source: Watson Institute for International and Public Affairs, Brown University. *Costs of war project*. http://watson.brown.edu/costsofwar/about.

Note
DOD – Department of Defense; GWOT – Global War on Terrorism; OCO – Overseas Contingency Operations. Dollars rounded to the nearest billion.

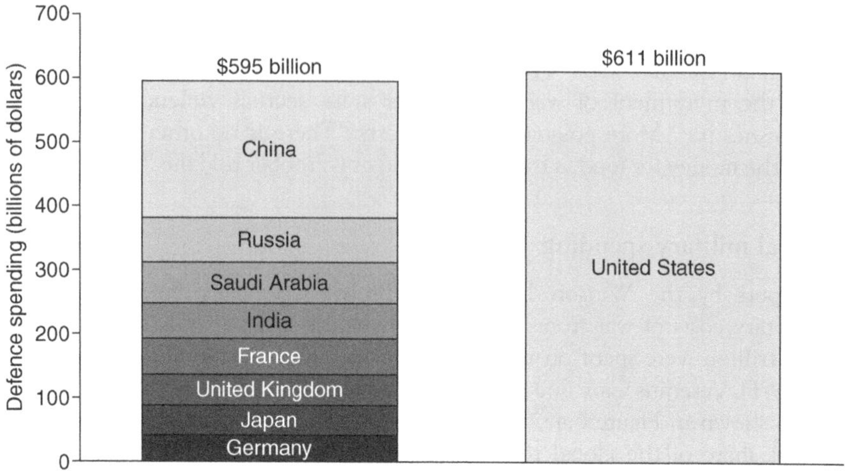

Figure 6.4 The United States spends more on defence than the next eight countries combined.

Source: Stockholm International Peace Research Institute. *Military expenditure database* (May 2018). Data are for 2017, compiled by Peter G Peterson Foundation, 2018 © Peter G Peterson Foundation. www.pgpf.org/chart-archive/0053_defense-comparison.

Note
Figures are in US dollars, converted from local currencies using market exchange rates.

expenditure by 5.9 per cent to US$69.2 billion. Western Europe in turn increased defence spending by 2.4 per cent.

At the seventieth session of the UN General Assembly in 2015 when the 2030 Agenda was adopted, including the 17 SDGs, then Secretary General Ban Ki-moon posed an important existential question: 'Why is it easier to find the money to destroy people and planet than it is to protect them?'.[26]

A similar refrain has been heard from the WEF.[27] Citing the 2016 *Global Peace Index* report, the WEF concluded 'that 2015 was a bad year for international peace and security'; there was 'a further deterioration in global peace based on historic trends' and 'the highest number of global battle deaths for 25 years, persistently high levels of terrorism, and the highest number of refugees and displaced people since World War II'.[28]

In a report, the Institute for Economics and Peace (IEP), 'a think tank dedicated to developing metrics to analyze peace and to quantify its economic value',[28] highlighted that *the economic impact of violence on the global economy was US$13.6 trillion (or 13.3 per cent of world GDP) in 2015 in terms of purchasing power parity (PPP)*. Moreover, this may be 'considered conservative as there are many items that have not been counted in the study due to the lack of available data'.[28] This is equivalent to US$1,876 PPP per annum, per person or US$5 *per day* for every person on the planet, or '11 times the size of global foreign direct investment'.[28]

In contrast, the amount spent on peace-building – the activities that aim to create peace in the long term – totalled US$6.8 billion or only 0.9 per cent of the economic losses from conflict![28]

The 'war-peace' spending ratios confirm that while human progress in the scientific and technological fields has been remarkable, especially since the mid-eighteenth century, our quest for civilised societies appears to be in crisis. This is confirmed by correspondent Martin Fletcher's visit to an exhibition, 'Syria: A Conflict Explored', at the Imperial War Museum North in Manchester, UK.[29] Fletcher has reported on the Syrian conflict over the last seven years – 'longer than the Second World War' – and recounts the tragedy that has ensued and which, he says, 'has numbed the West', perhaps because its 'constantly shifting dynamics have become far too complex to follow'.[29] Almost 'half a million men, women and children have been killed. Eleven million people – more than half the population – have been forced from their homes'.[29] Fletcher highlights that the war, which began from 'a popular uprising against an oppressive regime', turned 'into a multidimensional international proxy war'.[29] It pitted

> Sunnis against Shias, Iran against Saudi Arabia, the US against Russia and Turkey against the Kurds. It has given rise to the most ruthless jihadists the world has seen. It has triggered a refugee crisis that has fractured the European Union and fuelled the anxieties that underpinned Brexit and Donald Trump's election. Much of Syria has been destroyed, the ruins of the 21st century now joining the ruins of previous civilisations in that ancient country, and still the war rages on without the remotest prospect of peace in the foreseeable future.[29]

Fletcher's attention at the exhibition was drawn to 'harrowing photographs' taken by Sergey Ponomarev, the Pulitzer prizewinning photographer, 'which underscore the desperate human consequences of war':[29]

> Homeless boys playing in apocalyptic ruins, a man drinking from a hose amid an expanse of arid rubble, blindfolded prisoners standing against a prison wall, women nervously drinking coffee in a Damascus café. And, of course refugees in every guise – giving thanks as they reach Greek shores, huddled in blankets against the bitter cold, gathered around camp fires, walking up railway lines, pressing to cross borders, hopeful, lost, despairing as they trudge across Europe in scenes reminiscent of the aftermath of the Second World War.[29]

With over 3 million displaced in Yemen[30] and hundreds of thousands of Rohingya Muslims fleeing to Bangladesh,[31] alongside thousands of African migrants arriving on European doorsteps every year, humanitarian questions are many, the answers few. Achieving UN SDG 16 which calls for the promotion of 'just, peaceful and inclusive societies' through 'the provision of access to justice for all, and building effective, accountable institutions at all levels' seems increasingly remote.[32]

As illustrated in Figure 6.5,[33] while interstate wars leading to 'atomic annihilation of an entire city and inhabitants' are still possible, they may be improbable. In his book, *The Future of War: A History*,[34] Lawrence Freedman, one of Britain's foremost military thinkers, suggests future wars will likely 'be fought in urban environments', largely 'because of the mutually destructive power of nuclear weapons, international legal constraints and the declining appetite for violence of relatively prosperous societies'. Matthew Symond, a University College London researcher in the use of the digital humanities, who contributed to a special report, 'Why nuclear stability is under threat',[35] concurs with this analysis.

Data from 1946 to 2016 (see Table 6.2) show considerable decreases in extrastate and interstate conflicts but considerable and worrying increases in internationalised intrastate conflicts, with those in 2015 and 2016 involving foreign powers rising to 20 in 2015, dropping only slightly in 2016.

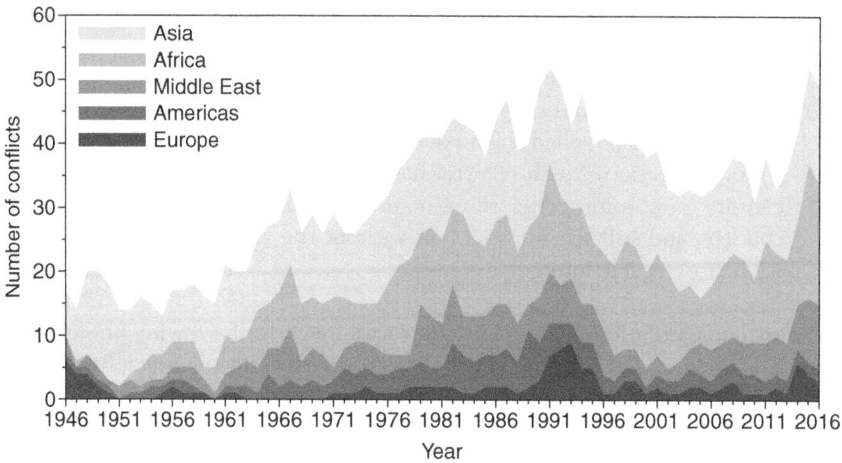

Figure 6.5 Armed conflict by region, 1946–2016.

Source: Department of Peace and Conflict Research, Uppsala University.

Table 6.2 Selected types of armed conflict, years 1946–2016

Year	*Extrastate*	*Interstate*	*Internationalised intrastate*	*Intrastate*
1946	5	2	2	8
1991	0	2	2	48
2014	0	1	13	28
2015	0	2	20	31
2016	0	2	18	29

Given the continuing number of countries currently involved in armed conflicts as well as potential others (e.g. North Korea), Freedman is right to acknowledge that there is now 'greater opportunity to misjudge intentions' and that this could lead to 'the human catastrophe that a nuclear war might unleash'.[34] As was the case during the Russia–US Cuban Missile Crisis in 1962, nations may still need to rely on preventing a nuclear war through 'the moral sympathy' of 'ideologically opposed' individuals.[34]

Threats on a larger scale remain, however. For example, Russia and China

> have an interest in challenging the American-sponsored international order, and both have recently shown that they are prepared to apply military force to defend what they see as their legitimate interests: Russia by annexing Crimea and destabilising Ukraine, and China by building militarised artificial islands and exerting force in disputes with regional neighbours in the South and East China Seas.[35]

A key difference to previous encounters in dealing with the US and other countries opposed to expansionism, says Symonds, is 'exploiting a "grey zone" between war and peace', applying 'calibration, leverage and ambiguity' and reaping 'either political or territorial gains normally associated with overt military aggression without tipping over the threshold into open warfare with a powerful adversary'.[35] One of their main fears could be that 'technological advances could allow America to threaten their nuclear arsenals without resorting to a nuclear first strike'.[35]

Symond's article underlines perhaps the biggest concern in the technology era (as discussed in Chapter 3) and relates to 'the rapid advances in artificial intelligence (AI) and deep learning' and how these 'will affect the way wars are fought, and perhaps even the way people think of war'.[35] The likelihood that these technologies will be able to 'create autonomous weapons systems that can make choices about killing humans independently of those who created or deployed them'[35] is high. This is especially so given, as history tells us, that some countries may not 'comply with the laws of armed conflict' – for example, 'a human must always be at least on the [decision-making] loop' – and 'may not be so scrupulous if fully autonomous systems are seen to confer military advantages'.[35]

UN adoption of a 'peace continuum'

It is for these reasons that UN General Assembly President, Miroslav Lajčák's groundbreaking recent announcement at a conference in Doha, Qatar, regarding the safety and security of the planet is extremely important.[36] Rather than focusing on the funding of wars, the President set out a new approach to conflict resolution recommending the adoption of a 'peace continuum'. The idea is to bring together 'under one umbrella' all the stages of conflict – 'prevention, mediation and peacebuilding'.[36] The President called on world leaders to 'voice their commitment to making financing for Sustaining Peace stronger, and more predictable'.[36] He also affirmed 'the importance of partnerships',[36] including 'within the UN system' where 'silos between humanitarian and development work must be torn down' and 'partnerships between the UN and other actors, including civil society, academia, regional and sub-regional organizations, regional development banks, and international financial institutions' should be strengthened.[36]

At the centre 'of all the sustaining peace efforts' are 'gender equality and women's empowerment'.[37] According to UN Secretary-General António Guterres, it is crucial that 'to enable women to play their essential role in preventing and resolving conflicts' they 'occupy leadership positions in the national, regional and global mechanisms for conflict resolution'.[37]

A fundamental challenge that threatens to undermine this UN peace initiative, however, relates to the world's current level of instability, distrust and uncertainty in respect of anything global. Questions about 'getting back to basics' – the building blocks of sustainability: ecological, social, geopolitical, financial and technological – are now more than evocative or wishful thinking. Water, energy, food security and climate increasingly require a search for pragmatic decisions to ensure the global survival of all species. Unfortunately, finding consensus around these survival issues is continually side-tracked and threatened by ongoing dangerous, human-centric, ideological, territorial and often self-centred battles at the highest levels – despite the fact that history has repeatedly demonstrated that there can be no winners in the long run and that genuine collaboration is the only way forward.

The Doomsday Clock: now *'two minutes to midnight'*

As mentioned in the Introduction, the Doomsday Clock, maintained by members of the *Bulletin of the Atomic Scientists*' Science and Security Board, has now been moved closer to midnight, from two and a half minutes to two minutes.[38]

Once again, the Science and Security Board recognises that 'Major nuclear actors are on the cusp of a new arms race, one that will be very expensive and will increase the likelihood of accidents and misperceptions' and 'urges leaders and citizens alike to redouble their efforts in committing to a path that advances the health and safety of the planet'.[39]

Science and Security Board recommendations include, inter alia, an 'insufficient response to climate change', in particular, the US government's 'unwise and ineffectual policies on climate change' and 'emerging technologies and global risk', especially 'the loss of public trust in political institutions, in the media, in science, and in facts themselves – a loss that the abuse of information technology has fostered'.[39] The Board highlights 'attempts to intervene in elections through sophisticated hacking operations and the spread of disinformation', both of which 'have threatened democracy, which relies on an informed electorate to reach reasonable decisions on public policy – including policy relating to nuclear weapons, climate change, and other global threats'.[39] The Board also suggests that 'internet companies such as Facebook and Google have been slow to adopt protocols to prevent misuse of their services and protect citizens from manipulation' and calls for the international community to 'establish new measures that discourage and penalize all cross-border subversions of democracy'.[39]

In its concluding comments, the Science and Security Board reminds us that: 'This is a dangerous time, but the danger is of our own making. Humankind has invented the implements of apocalypse; so can it invent the methods of controlling and eventually eliminating them.'[39] In short, while 'the failure of world leaders to address the largest threats to humanity's future is lamentable ... that failure can be reversed'[39] – less so, unquestionably, should the Doomsday Clock be pushed forward another 30 seconds.

> My name is Ozymandias, King of Kings;
> Look on my Works, ye Mighty, and despair!
> Nothing beside remains. Round the decay
> Of that colossal Wreck, boundless and bare
> The lone and level sands stretch far away.
> (Shelley's *Poetry and Prose*)[40]

References

1 Tolstoy L. (trans. Ann Dunnigan). *War and Peace*. London: Penguin Books; 1968.

2 The Quarterly Conversation. *War and Peace by Leo Tolstoy* [review]. http://quarterly-conversation.com/war-and-peace-by-leo-tolstoy-review (accessed 14 July 2015).

3 Sorcinelli G. *Holmes Wilson and the Timeless Themes of 'War and Peace' by Leo Tolstoy*. www.linkedin.com/pulse/holmes-wilson-timeless-themes-war-peace-leo-tolstoy-gino-sorcinelli/ (accessed 20 August 2016).

4 Daley J. We're repeating the mistakes that pitched the world into disaster last century. But there's still hope. *Daily Telegraph*, 8 July 2017. www.telegraph.co.uk/news/2017/07/08/repeating-mistakes-pitched-world-disaster-last-century-still/ (accessed 10 August 2017).

5 Freedom House. *Champions for freedom*. https://freedomhouse.org/our-work (accessed 15 January 2018).

6 Abramowitz MJ. *Democracy in crisis*. https//freedomhouse.org/report/freedom-world/freedom-world-2018 (accessed 10 February 2018).

7 European Council on Foreign Relations. *A normative crisis: How to protect democratic values in Europe*. www.ecfr.eu/article/commentary_how_to_protect_democratic_values_in_europe_7227 (accessed 19 January 2018).

8 Human Rights Watch. *World report 2018*. www.hrw.org/world-report/2018 (accessed 24 February 2018).

9 *The Economist. Democracy continues its disturbing retreat*. www.economist.com/blogs/graphicdetail/2018/01/daily-chart-21 (accessed 10 February 2018).

10 Gerling RW. *Cyber attacks on free elections*. www.mpg.de/11357138/W001_Viewpoint_010-015.pdf (accessed 19 November 2017).

11 Burleigh M. There's never been a worse moment in history for a parliament frozen in fear. *Mail on Sunday*, 5 February 2017.

12 Ferguson N. The rise of intelligence threatens to make the world more mystifying. *The Sunday Times*, 11 February 2018: 25.
13 Harari YN. *Sapiens: A Brief History of Humankind.* London: Harvill Secker (Random House); 2014.
14 Ritholtz B. *Global deaths in conflicts since 1400*. http://ritholtz.com/2015/06/global-deaths-in-conflicts-since-the-year-1400/ (accessed 27 September 2017).
15 International Physicians for the Prevention of Nuclear War. *Body Count: casualty figures after 10 years of the 'war on terror' – Iraq, Afghanistan, Pakistan*. www.ippnw.de/commonFiles/pdfs/Frieden/Body_Count_first_international_edition_2015_final.pdf (accessed 20 July 2018).
16 Engelhardt T. *Mapping a world from hell*, www.tomdispatch.com/blog/176369/tomgram%3A_engelhardt%2C_seeing_our_wars_for_the_first_time/ (accessed 10 January 2018).
17 Engelhardt T, Fraser SS. *American empire project – About the project*. http://Americanempireproject.com/about/ (accessed 12 January 2018).
18 Freedman I. Can there be peace with honor in Afghanistan? *Foreign Policy Magazine*, 26 June 2017. http://foreignpolicy.com/2017/06/26/all-thats-left-in-afghanistan-is-defeat-with-honor/ (accessed 10 November 2017).
19 Watson Institute (Brown University). *Costs of war project*. http://watson.brown.edu/costsofwar/about (accessed 12 January 2018).
20 Taylor A. 149,000 people have died in war in Afghanistan and Pakistan since 2001, report says. *Washington Post*, 3 June 2015.
21 The Associated Press. A timeline of U.S. troop levels in Afghanistan since 2001. *Military Times*, 6 July 2016. www.militarytimes.com/news/your-military/2016/07/06/a-timeline-of-u-s-troop-levels-in-afghanistan-since-2001/ (accessed 12 November 2016).
22 Westcott B. *Afghanistan: 16 years, thousands dead and no clear end in sight*. https://edition.cnn.com/2017/08/21/asia/afghanistan-war-explainer/index.html (accessed 12 November 2017).
23 Loyd A. Afghanistan: The war that never ends. *The Times Magazine*, 11 November 2017.
24 Crawford NC. *United States budgetary costs of post-9/11 wars through FY2018*. http://watson.brown.edu/costsofwar/files/cow/imce/papers/2017/Costs%20of%20U.S.%20Post-9_11%20NC%20Crawford%20FINAL%20.pdf (accessed 15 January 2018).
25 Peter G Peterson Foundation. *The United States spends more on defense than the other eight nations combined*. www.pgpf.org/chart-archive/0053_defense-comparison (accessed 16 January 2018).
26 Deen T. 'Why is it easier to find money to destroy people than protect them?' asks U.N. chief. *Inter Press Service*, 28 September 2015. www.ipsnews.net/2015/09/why-is-it-easier-to-find-money-to-destroy-people-than-protect-them-asks-u-n-chief/ (accessed 17 November 2016).
27 World Economic Forum. *War costs us $13.6 trillion. So why do we spend so little on peace?* www.weforum.org/agenda/2016/06/the-world-continues-to-spend-enormous-amounts-on-violence-and-little-on-building-peace/ (accessed 10 June 2016).
28 Institute for Economics and Peace. *Global peace index 2016*. http://economicsandpeace.org/wp-content/uploads/2016/06/GPI-2016-Report_2.pdf (accessed 19 November 2018).
29 Fletcher M. The reality of Syria in all its mundane horror. *The Times*, 16 February 2016: 12–13.
30 United Nations High Commissioner for Refugees. *3 million displaced in Yemen*. www.unhcr.org/afr/news/press/2017/2/58ac0b170/yemens-brutal-conflict-pushing-million-displaced-return-danger-joint-unhcr.html (accessed 4 October 2017).
31 Reuters staff. Persecuted Rohingya Muslims flee violence in Myanmar. *Reuters*, 19 December 2017. www.reuters.com/article/us-global-poy-myanmar-rohingya/persecuted-rohingya-muslims-flee-violence-in-myanmar-iduskbn1ed2cg (accessed 9 January 2018).

32 United Nations. *Sustainable Development Goal 16*. https://sustainabledevelopment.un.org/sdg16 (accessed 15 November 2017).
33 Department of Peace and Conflict Research, Uppsala University. *Number of armed conflicts by type*. www.pcr.uu.se/research/ucdp/charts-graphs-and-maps/#tocjump_9236458892964821_1 (accessed 14 December 2017).
34 Freedmam L. *The Future of War: A History*. New York: Hachette Book Group; 2017.
35 Symonds M. The future of war. *The Economist*, 25 January 2018. www.economist.com/news/special-report/21735477-war-still-contest-wills-technology-and-geopolitical-competition-are-changing (accessed 27 January 2018).
36 United Nations. In Doha, General Assembly President spotlights UN 'sustaining peace' agenda. *UN News*, 18 January 2018. https://news.un.org/en/story/2018/01/1000711 (accessed 20 July 2018).
37 Lebada AM. UN Secretary-General briefs governments on peacebuilding, sustaining peace. *International Institute for Sustainable Development*, 13 March 2018. http://sdg.iisd.org/news/un-secretary-general-briefs-governments-on-peacebuilding-sustaining-peace/ (accessed 15 March 2018).
38 Bulletin of the Atomic Scientists staff. It is now 2 minutes to midnight. *Bulletin of the Atomic Scientists*, 25 January 2018. https://thebulletin.org/press-release/it-now-2-minutes-midnight11464 (accessed 29 January 2018).
39 Bronson R. It is now two minutes to midnight. Doomsday clock statement. *Bulletin of the Atomic Scientists*, 25 January 2018. https://thebulletin.org/2018-doomsday-clock-statement (accessed 8 February 2018).
40 Shelley PB. *Shelley's Poetry and Prose: Authoritative Texts, Criticism*. 2nd ed. London: Norton; 2002.

7 The world in 2050

Megatrends and social impacts

Evolutionary stages of society

In an article published several years ago, sociologist Keeley Knowles suggested that many of the problems in the more economically advanced world could be traced (see Box 7.1) to the beginning of the industrial era which 'saw an unparalleled shift in familial life, leading to a loss of traditional values'.[1]

Box 7.1 Evolutionary stages of society – c.12,000 BCE to early twenty-first century

Until about 12,000 BCE

Families lived as hunter–gatherers – 'kinship based with the family traditionally taking responsibility for gathering and distributing food, educating, and bestowing essential survival skills on their children'.[1]

Around 5,000 years ago in the Middle East

Hunter–gatherer societies transformed into an agrarian society, characterised by large-scale farming 'and additional scientific advances, for example the use of the wheel, and irrigation, thereby transforming 'all areas of social life'.[1]

Mid-eighteenth century

The beginning of the industrial era 'saw an unparalleled shift in familial life, leading to a loss of traditional values, changes in working patterns and the kinds of work available, and urbanisation – overall, a more impersonal, individualistic way of life'.[1]

Early twenty-first century

Today's world – the 'post-industrial, information-based society'[2] – Knowles observed, may be best understood 'through the imagery of global multicultural companies such as McDonald's': the process by which 'the principles of the fast-food restaurant are coming to dominate more and more sectors of American society as well as the rest of the world'.[1]

'McDonaldization', a term coined by George Ritzer in his 1993 book, *The McDonaldization of Society*,[2] 'continues to stand as one of the pillars of modern day sociological thought'.[2] Its principles – 'efficiency, calculability, predictability, control'[2] – have also been associated with such words as 'Amazonization', which refers to the values and business practices of the Amazon corporation, the world's top online retailer. These characteristics of what are perceived to be well-functioning organisations are now also applied to many other sectors including manufacturing, banking, education, healthcare and online shopping, to name a few. The four principles cited above are defined as follows:

- *efficiency* – 'the choosing of means to reach a specific end rapidly, with the least amount of cost or effort';
- *calculability* – an emphasis on 'things that can be calculated, counted, quantified. Quantification refers to a tendency to emphasize quantity rather than quality';
- *predictability* – the attempt 'to structure our environment so that surprise and differentness do not encroach upon our sensibilities'; and
- *control* – the 'Substitution of Nonhuman for Human Technology'.[2]

To these four attributes George Ritzer added a fifth dimension, 'irrationality of rationality', meaning that 'rational systems are unreasonable systems ... they deny the basic humanity, the human reason, of the people who work within or are served by them'.[2]

Now extending to most other global businesses and services (including higher education) – such as Ikea, Walmart, Apple, Disney World and Samsung – these operating principles and practices (put in the context of customer transactions) exist, according to Jonathan Kittow, director at a supply chain company, primarily 'to create a 'perfect order', (algorithm?) 'which performs as expected at 100% throughout each step or intervention within an order's life – from creation to payment'.[3] Put more technically, 'The measurement of a "perfect order" ... is about how much deviation there is, during this process that makes it less than 100% perfect.'[3]

In practical terms, this approach demands that online orders are accepted, confirmed and delivered within 24 hours, 'suppliers ... accept and confirm orders for delivery within seconds and ensure that consumers receive their products – often within 24 hours' and that there is 'an electronic audit trail, all the way from order to payment (the order-to-cash part of the supply chain)'.[3]

'McDonaldization' and 'Amazonization': illusions of efficiency and economy?

In a student note on 'The Irrationality of Rationality' to accompany Chapter 7 of Ritzer's book, *The McDonaldization of Society*, Dr John Mahoney argues that 'Whole industries are now in the business of producing and marketing unreality'.[4] Mahoney references George Ritzer's concluding comments related to the food industry:

> Thus is the argument of this text that contrary to McDonald's propaganda, and the widespread belief in it, fast-food restaurants and their rational clones are not reasonable, or even truly rational, systems. They spawn problems in the health of their customers and the well-being of the environment; they are dehumanizing and therefore, unreasonable; and they often lead to the opposite of what they are supposed to create, for example, leading to inefficiency rather than to 'increased efficiency' for it is within the context of the bureaucracy that dehumanization really occurs.[4]

Professor Dorothy Wills at California Polytechnic State University is also concerned about the consequences of depersonalization in our everyday lives. She contends that 'The idea of efficiency is specific to the interests of the industry or business, but is typically advertised as a benefit to the customer'[5] (e.g. ATMs, self-serve shopping). Wills highlights that: 'The interesting element here is that the customer often ends up doing the work that previously was done for them.'[5] She reminds us that while there have been 'many benefits and conveniences that are related to this process of McDonaldization: variety, round-the-clock banking and shopping, and often speedier service', she agrees with George Ritzel that 'there is a certain sense that these rational systems tend to turn in on themselves, to lead to irrational outcomes'.[5]

Moreover, Wills maintains that

> rational system don't save us money; we might spend less, but we do more work. The food we eat is often less nourishing, loaded with stabilizers and flavor enhancers, fats, salt and sugar. This contributes to the health problems of our society, a definitely "antihuman" component. As our children grow up within these systems, they develop habits which insure our increasing dependency upon the systems. The packaging used in [the] fast food industry pollutes the environment. And the family: part of its solidarity and integrity was centered around the family meal: The communal meal is our primary ritual for encouraging the family to gather together every day. If it is lost to us, we shall have to invent new ways to be a family. It is worth considering whether the shared joy that food can provide is worth giving up.[5]

Furthermore, Wills reasons that in a society where 'everything is pre-packaged, pre-measured, automatically controlled', it is becoming apparent that: 'The human employee is not required to think, just follow the instructions and push a button now and then'.[5] The implications for 'who we are' and 'how we interact' as well as for education and training are profound as we will likely be increasingly 'defined by our dependence upon and subordination to the machine' as 'the skills and capabilities of the human actor are quickly becoming things of the past'.[5]

In a background paper prepared for her students,[5] Wills raises several fundamental questions that should concern us all:

- How long will it be before these rational systems evolve beyond the control of people?
- What happens when the people who control the systems succumb to being controlled?
- What is the possibility of a small number of individuals exercising tremendous control over the people dependent upon the systems?

As things stand, and as highlighted in previous chapters, we are at a historic turning-point when decisions – social, economic, geopolitical, environmental – we make today could determine the future well-being of our own species as well as other species with whom we share the world. Along with the challenges discussed earlier – climate change and biodiversity, urbanisation and demographics, modernity, technological advances in cognitive computing and the evolving nature of capitalism, longer term global economic trends will increasingly play a pivotal role in identifying sustainability issues and, in particular, how these might impact on society and the planet.

The world in 2050: shifting global economic power?

In their report, 'Leading in changing times: making a difference',[6] PricewaterhouseCoopers (PwC), a multinational professional services network headquartered in London, predicts that by 2050 global economic power will shift from the G7 countries – US, UK, France, Germany, Japan, Canada, Italy – to the E7 nations (group of emerging market economies) – China, India, Indonesia, Brazil, Russia, Mexico, Turkey – with emerging markets topping the world economies (see Figure 7.1).

	2016	2050	
China	1	1	*China*
United States	2	2	*India*
India	3	3	*United States*
Japan	4	4	*Indonesia*
Germany	5	5	*Brazil*
Russia	6	6	*Russia*
Brazil	7	7	*Mexico*
Indonesia	8	8	*Japan*
United Kingdom	9	9	*Germany*
France	10	10	*United Kingdom*

□ E7 economies ■ G7 economies

Figure 7.1 Projected GDP rankings (at PPPs) – 2016 vs 2050.

Sources: International Monetary Fund for 2016 estimates; PwC for projections to 2050.

Note:
Gross domestic product (GDP) at purchasing power parity (PPP) adjusts for price level differences across countries and provides a better measure of the volume of goods and services produced in an economy.

An important caveat upon which this re-alignment would be based is that the 'emerging economies need to enhance their institutions and their infrastructure significantly if they are to realise their long-term growth potential'.[6] Other conditions for success include avoiding 'a slide back into protectionism … ensuring that the potential benefits of globalisation are shared more equally across society … [and] developing new green technologies to ensure that long-term global growth is environmentally sustainable.'[6]

Barring geopolitical and other factors, the predicted top 32 economies and their comparative GDP rankings for 2016, 2030 and 2050 are shown in Figure 7.2. The greatest GDP decreases lie in store for Europe – Italy (from 12 to 21), Spain (from 16 to 26), Poland (from 23 to 30) and the Netherlands (from 26 to 32). Australia is expected to drop from nineteenth to twenty-eighth place.

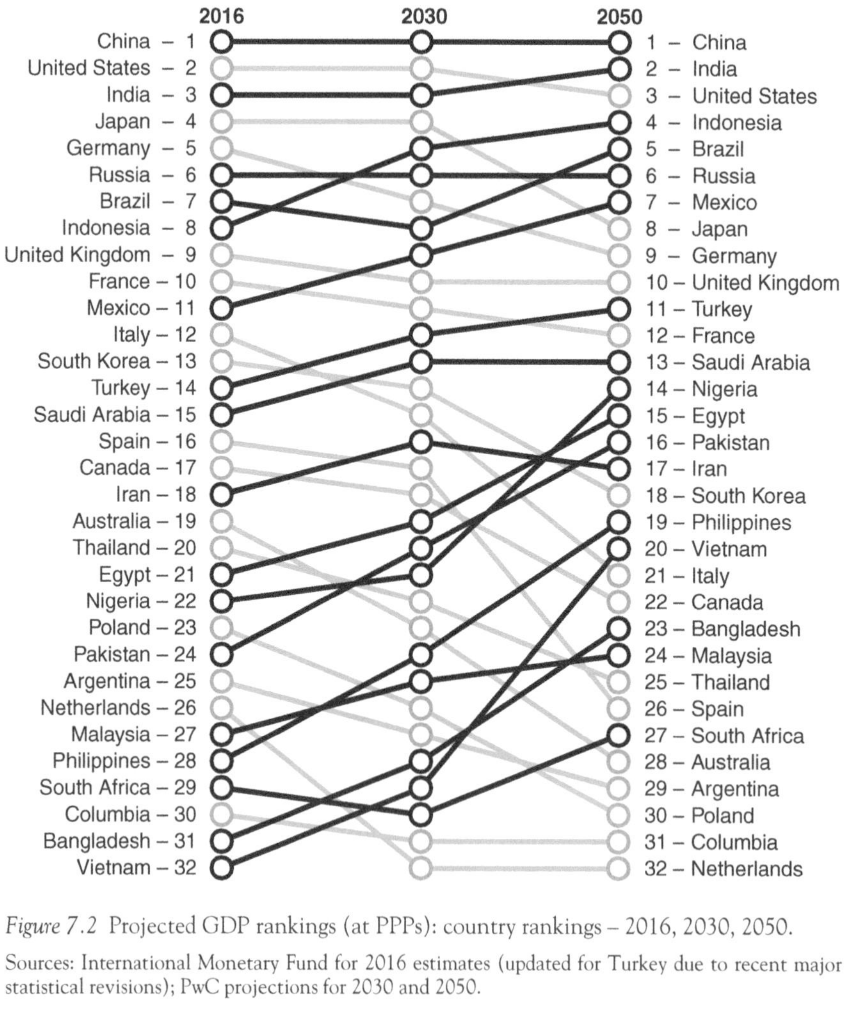

Figure 7.2 Projected GDP rankings (at PPPs): country rankings – 2016, 2030, 2050.

Sources: International Monetary Fund for 2016 estimates (updated for Turkey due to recent major statistical revisions); PwC projections for 2030 and 2050.

Indonesia may rise from eighth to fourth place, Mexico from eleventh to seventh, the Philippines from twenty-eighth to nineteenth, Pakistan from twenty-fourth to sixteenth and Vietnam from thirty-second to twentieth place. The US and UK are expected to drop one place each from their position in 2016 – the US from second to third and the UK from ninth to tenth.

Societal impacts of automation in developing and advanced economies

The impact of technology on the workforce, as discussed in Chapter 3, remains equivocal. Based on an analysis of 200,000 jobs in 29 nations 'to explore the economic benefits and challenges posed by automation', PwC projections are that 'AI, robotics and other forms of smart automation have the potential to bring great economic benefits, contributing up to $15 trillion to global GDP by 2030'[7] but there are also concerns that many existing jobs could be displaced.

As shown in Figure 7.3, three waves of automation are anticipated. During the first wave, relatively low displacement of existing jobs is expected perhaps only around 3 per cent by the early 2020s. However, job displacement could increase in later waves as technologies mature and are rolled out across the economy in increasingly autonomous form. By the mid-2030s, up to 30 per cent of jobs could be automatable, with slightly more men being affected in the long run as autonomous vehicles and other machines replace many manual tasks where their share of employment is higher.

During the first and second waves, however, women could be at greater risk from automation due to their higher representation in clerical and other administrative functions (see Figure 7.3). These estimates are median values across 29 countries, with the UK being very close to the average. Long-term automation

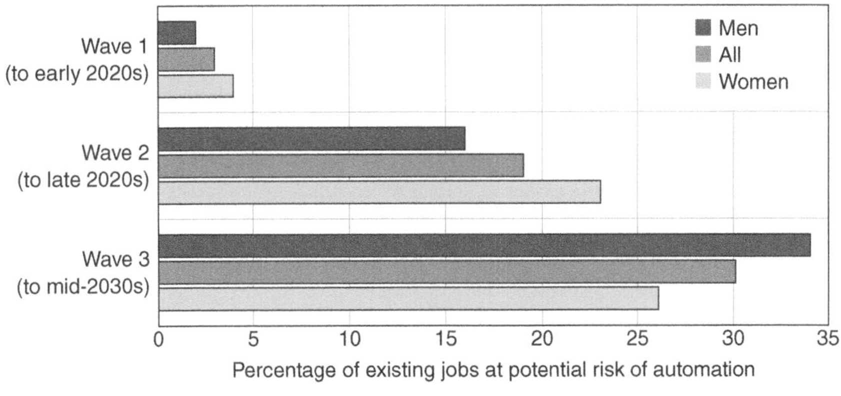

Figure 7.3 Risks of automation.

Source: PwC estimates based on OECD Programme for the International Assessment of Adult Competencies (PIAAC) data (median values for 29 countries).

could be lower at only around 20–25 per cent in Asian and Nordic countries, but could be higher at over 40 per cent in some Eastern European countries.[7]

PwC observes that the impact of these developments 'will vary substantially across different regions',[7] especially in light of the fact that Africa's population – the fastest growing – is set to double by 2050. Europe's population is expected to shrink while 'fertility in Latin America will remain higher than mortality'.[7] And while 'the average age in Japan in 2050 will be 53 – in Nigeria it will be 23'.[7]

Overall, it is projected that the world's population will rise 'by more than 1 billion, bringing the total to over eight billion'[7] and that '97% of this population growth will come from emerging or developing countries'.[7] In addition, and equally significantly, PwC posits that people in all regions are living longer and having fewer children. The result is that the fastest growing segment of the population will be the over 65s – there will be 390 million more of them in 2030 than in 2015.[7]

PwC 'Megatrends'[7] (see Box 7.2) confirm the findings discussed in Part I.

Box 7.2 Global megatrends to 2030

Rapid urbanisation

Today, more than half the world's population live in urban areas and almost all of the new growth will take place in lesser known medium-sized cities in developing countries.

Climate change and resource scarcity

As the world becomes more populous, urbanised and prosperous, demand for energy, food and water will rise. But the Earth has a finite amount of natural resources to satisfy this demand.

Shift in global economic power

Some emerging economies that were growing rapidly are now in recession. Commodity prices have played a considerable role in sending these economies into reverse.

Demographic and social change

By 2030 the world's population is projected to rise by more than 1 billion. Equally significantly, people are living longer and having fewer children.

Technological breakthroughs

The digital revolution has no boundaries or borders. It is changing behaviour and expectations as much as the tools used to deliver new services and experiences.

Challenges for policymakers – achieving sustainable growth

PwC cautions that in order to address the challenges facing the globe 'all countries will need to implement bold policies to cope with these demographic changes',[7] in particular those facing an ageing population such as North America and Europe, as well as much of Asia and Latin America. As discussed in Chapter 5, 'supporting an ageing population will require greater participation in the labour force from women and the elderly, and possibly also higher levels of immigration.'[7] First, structural reforms with a view to improving 'macroeconomic stability, diversify[ing] their economies away from undue reliance on natural resources (where this is currently the case), and develop[ing] more effective political and legal institutions'[6] are essential because of exponential population growth and, as mentioned above, negative population growth in many developed nations. Second, growing global inequality needs to be addressed in a world where the '8 richest people now own as much wealth as the poorest *half* of the world's population',[6] thus reminding nations that new 'bold policies' are also needed to address the 'vicious cycle' that has developed between corruption, 'unequal distribution of power' and 'unequal distribution of wealth'.[8]

Tackling corruption worldwide

Perhaps at no other time in living memory has there been a greater need for innovative, workable and sustainable approaches to ensuring 'a globalisation that works for all', 'sharing the spoils of growth' or 'redistributing some of the gains from trade and technology'[9] while tackling corruption and organised crime on a grand scale. Misha Glenny, investigative reporter and writer of the hit series *McMafia*, argues that organised crime, which in recent decades has been 'globalised along with capitalism and now accounts for as much as 15 percent of the world's economy', is a greater threat to the world than terrorism – both economically and in terms of the costs to human life.[10] He cites Enron, Fifa, drug wars in Mexico and Brazil, where tens of thousands are killed every year, and 'people who are trafficked from Nepal into the Gulf States to work on building sites' and 'die in their hundreds every year' as examples. Profit is the main motivation and 'weakening democratic institutions and processes' is a key factor evidenced by chaos in the Middle East, the American constitutional system, Brexit and the European Union.

International and national responses to this growing problem have not been in vain but are not enough as shown in Figure 7.4.[11] The 2016 Corruption Perception Index, produced annually by Transparency International (TI), highlights that

> Over two-thirds of the 176 countries and territories … fall below the midpoint of the [TI] scale of 0 (highly corrupt) to 100 (very clean). The global average score is a paltry 43, indicating endemic corruption in a country's public sector.[11]

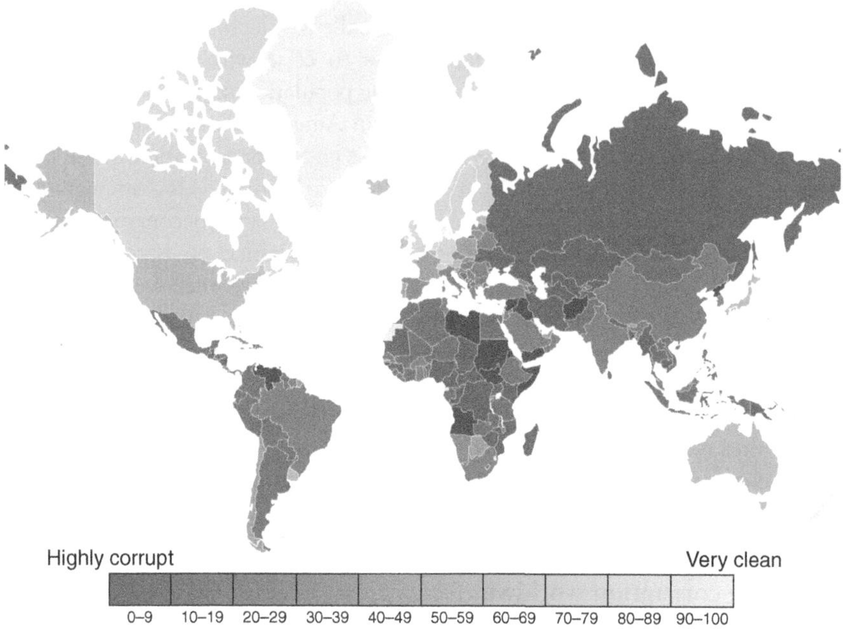

Figure 7.4 Corruption perceptions index 2016.

Source: Transparency International. www.transparency.org/news/feature/corruption_perceptions_index_2016.

Top-scoring countries are far outnumbered by corrupt or highly corrupt countries, where citizens face the tangible impact of corruption on a daily basis.

Government responses have varied from Chinese President Xi Jinping's anti-corruption drive[12] and President Ramaphosa's war on corruption in South Africa[13] to World Bank Group global approaches. For example, in 1996 the World Bank introduced 'Worldwide Governance Indicators', which measure the quality of governance in around 200 countries, based on over 30 individual data sources.[14] An interactive data access tool[15] allows users to select different views of the data (including, for example, graph and table views) and combine/compare data sets across nations (e.g., accountability, rule of law and corruption). As an example, Figure 7.5 combines and compares three indicators – Political Stability and Lack of Violence, Rule of Law, and Control of Corruption across Low Income, Lower Middle Income and Upper Middle Income countries. The data has remained relatively stable overall.

The World Bank Group: combatting corruption

Tackling corruption remains one of the World Bank Group's main priorities as it seeks to achieve 'its twin goals of ending extreme poverty by 2030 and boosting

World Governance Indicators: 6 Dimensional Framework: World Bank

- Voice and Accountability
- Political Stability and Lack of Violence
- Government Effectiveness
- Regulatory Quality
- Rule of Law
- Control of Corruption

Figure 7.5 World Bank governance indicators.

Source: Kaufmann D, Kraay A, Mastruzzi M. *The worldwide governance indicators* (2010).

shared prosperity for the poorest 40 percent of people in developing countries'.[16] Empirical studies have confirmed that 'an estimated $1.5 trillion' are paid 'in bribes each year … about 2% of global GDP – and 10 times the value of overseas development assistance'.[16] Most affected are the poor who pay 'the highest percentage of their income in bribes' which prevents many 'from accessing health services', thus impacting on 'health outcomes, such as infant mortality'.[16] Corruption also 'contributes to higher-order crimes' through 'illicit financial flows (IFFs) … [and] often finds its way across borders to fund drug and human trafficking', thereby eroding 'the social contract between citizens and the state' and harming foreign investment.[16]

In terms of strategies, the World Bank Group works with politicians, senior government officials, the private sector, citizens, communities and civil society organisations. Advanced technology 'to capture, analyze, and share information in order to prevent, detect, and sanction corrupt behaviour' is also increasingly used.[16] The World Bank Group also 'works at the country, regional and global level to build capable, transparent, and accountable institutions and design and implement anticorruption programs'.[16]

The Chartered Global Management Accountant 'Anti-corruption landscape 2017' report

In keeping with World Bank Group initiatives and taking responsibility 'to prevent corruption and improve fair and competitive conditions for business', Chartered Global Management Accountant (CGMA) has produced a joint update report (with Transparency International) which offers 'an overview of global corruption, outlines the legislative landscape in the UK and the US, describes how international measures to combat corruption have intensified and highlights the latest anti-corruption developments in a number of emerging markets'.[17]

The report's eight key anti-corruption procedures[17] are summarised in Box 7.3. This may be a useful self-assessment tool (e.g. 'To what extent has the company committed to…?') for monitoring the impact of existing corporate anti-corruption approaches.

Box 7.3 Corporate anti-corruption procedures

- **Top-level commitment**: The Board of Trustees or equivalent governing body should commit to, and oversee, the zero-tolerance policy and anti-bribery and anti-corruption programme, demonstrating visible and active commitment to its implementation.
- **Risk assessment**: Undertake regular bribery and corruption risk assessment that underpins the programme.
- **Effective anti-bribery policies and procedures**: Design and implement anti-bribery and anti-corruption policies and procedures that are effective and proportionate to the organisation's risks, circumstances and culture.
- **Incentives**: Ensure that the company's approach to incentives is aligned with and does not inadvertently undermine its anti-bribery and anti-corruption commitment.
- **Due diligence and procedures for third parties**: Carry out reasonable and proportionate due diligence on potential associates before entering into contracts with them, and put in place procedures for managing the associated risks on an ongoing basis.
- **Communication and training**: Build awareness and understanding of your anti-bribery and anti-corruption programme among the Board, employees, volunteers and relevant stakeholders through communication and appropriate training. Report publicly on the anti-bribery and anti-corruption measures.
- **Whistleblowing:** Encourage the reporting of bribery and corruption incidents, and provide secure and accessible whistleblowing and advice channels for use by all employees. Create a 'speak up' culture where employees know that their concerns will be listened to and acted upon, without risk of retaliation.
- **Monitoring and evaluation**: Monitor the implementation and effectiveness of the anti-bribery and anti-corruption programme. The results of monitoring should be reviewed regularly by the Board and guide improvements to the programme as necessary.

Democracy and internet freedom

The main thesis of Steven Pinker's book *Enlightenment Now*[18] is the supremacy of reason (an ideology of scientism?) – in particular that reason can replace religion. In contrast, an article by journalist Juliet Samuel on the threat the information age poses to democracy may come closer to identifying the fundamental limitations of relying solely on reason in the age in which we are living[19] (the 'Intelligence Era?') – where data and bits of information are beginning to reign supreme. Samuel's argument is that 'ideas of truth and scientific reason', originating from the time of the eighteenth-century Enlightenment, have now been taken over by new media which presents us with bits of information, not learning. Coupled with 'the loss of religion, family and the erosion of the nation' we have been 'left', she feels, 'adrift' to the point that 'the world has never seemed so meaningless'.[19] There are several consequences: 'Perspectives and "theories of thought" have replaced facts as arbiters of righteousness' – with 'fake news' exemplifying 'the application of this academic phenomenon on a mass scale'.[19]

At the present time, when societal 'loss of meaning brought by technology' and the need to develop 'a sense of belonging as well as freedom'[19] run deep, it may also be a propitious historical moment for reinventing democracy and confronting corruption – often a by-product of ineffective democratic institutions and processes.

For the late Kofi Annan, former Secretary-General of the UN and Nobel Peace laureate, 'Social media could be just the start of a slippery slope leading to an Orwellian world controlled by Big Data Brother, accelerated by convergence with the sensors in our devices and rapid advances in artificial intelligence.'[20] His concerns may remind us of the Dataist dogma world Yuval Harari pictures in *Homo Deus*,[21] cited in Part I of this book. Reflecting on the days when the internet and social media were 'hailed for creating new opportunities to spread democracy and freedom' (e.g., the Arab world in 2011), Annan observed that 'authoritarian regimes soon began cracking down on Internet Freedom' and 'most social media-enabled popular uprisings failed for want of effective leadership and traditional political and military organisations retained the upper hand'.[20]

Since that time, Annan argued, 'political systems have been severely challenged' and, like the 'printing press, radio and television', may benefit from 'rules of transparency, accountability and taxation as conventional media'.[20] These appear to be pressing given that 'Russia covertly used social media to influence electoral outcomes in the Ukraine, France, Germany, and most famously, in the United States'.[20] Further, Annan emphasised that 'even the most technologically advanced countries cannot protect the integrity of the electoral process', and commented that 'one can imagine the challenge facing countries with less know-how'.[20]

A recurring theme in the humanist vs tech literature is the extent to which Big Data and social media will be used globally as a source for good or evil. And, if the latter, what options does society have in respecting the privacy of data ownership, use of data and maximising accountability? To answer some of these questions, a new commission will be established under the auspices of the Kofi Annan Foundation 'with the masterminds of social media and information technology as well as political leaders – to help ... address these crucial new issues'.[20] The intended outcome is to produce recommendations that 'will ... reconcile the disruptive tensions created between technological advances and one of humanity's greatest achievements: democracy'.[20]

There is no question, as Michael Burleigh metaphorically confirmed, that 'the tectonic plates of world power are shifting'.[21] China's US$1 trillion 'One Belt, One Road' master plan[22,23] is 'already funding rail and sea links that will eventually connect Europe with China's rich industrial heartlands'.[21] 'Strategically', Burleigh observes, 'China thinks in the long term, we do not.'[21]

Columnist Janet Daley, identifies another important incongruity between past and present, arguing that 'the important struggle of our time' lies not 'between protectionism and free trade, or isolationism and engagement with allies' but 'between reason and unreason'.[24] Daley's main concerns are twofold. First, unlike 'modern democracy', which was spawned by the Age of Reason, the times we are living through seem to be dominated by 'unreason'; second, given the global polity, Daley asks, 'How long can the institutions and the liberty they were designed to protect, survive?'.[24]

Decision-making in the twenty-first century: pairing reason with heuristics, emotion and sustainability

In a *Quartz* article,[25] Olivia Goldhill, who writes on philosophy, psychology and neuroscience, draws on past research by eminent cognitive psychologists Daniel Kahneman (winner of the 2002 Nobel Prize in Economics) and Amos Tversky, who explored errors of thinking or 'cognitive biases'.[26] In their collaborative research they demonstrated 'that humans systematically make choices that defy clear logic',[25] a phenomenon that is regaining attention in terms of consumer behaviour.[27] One of the difficulties with reason, Goldhill affirms, is that 'we live in a world of deep uncertainty' where 'neat logic simply isn't a good guide' and where 'data-based decisions' do not 'inoculate against irrationality or prejudice'.[25] She observes that, rather than 'relying on complex statistics to make choices', we 'tend to make decisions according to instinct' – likely a by-product of our evolution – although these can 'lead us to bias or racist assumptions'.[25]

Emotions and heuristics ('mental shortcuts') play a central role in decision-making, Goldhill asserts, especially considering 'that *positive emotions* are often the ultimate ends of our decisions [emphasis added]'[25] – possibly confirmed by recent referenda and election results. Goldhill also cites cognitive scientist Hugo Mercier who reminds us that 'facts ... don't tell you anything ... it's only paired with preferences, desires, with whatever gives you pleasure or pain, that can guide your behavior'.[25] She highlights that 'our community of knowledge is absolutely critical to functioning' and that we cannot 'do anything alone' especially as 'technology gets more complex'.[25]

Goldhill's emphasis on the need to continue to rely on each other and her concluding comment that 'overall, we're still far better suited to the real world than the most perfectly logical thinking machine'[25] may bode well – for the world in 2050 and, hopefully, beyond.

It is likely that the longevity of 'institutions and liberty', both under threat,[24] could be optimised considerably *if* thinking, doing *and* feeling processes were better balanced and integrated across society, including the preparation of learners in the twenty-first century. This would be especially so if their learning experiences embraced a better understanding of the interdependence of the environment and all life,[28] a key dimension for ensuring the sustainability of people and planet espoused by the UN-2030 SDGs.[29]

References

1 Knowles K. *What is 'modernity' and why have sociologists been so interested in it?* www.essex.ac.uk/sociology/documents/pdf/ug_journal/vol.1/KeeleyKnowles_SC111_2008.pdf (accessed 15 July 2018).
2 Ritzer G. *The McDonaldization of Society: An Investigation into the Changing Character of Contemporary Social Life*. 8th ed. California: Sage Publications; 2015.
3 Kittow J. *What's the perfect order and why is it relevant now?* https://blog.omprompt.com/whats-the-perfect-order-and-why-is-it-relevant-now (accessed 20 February 2017).
4 Mahoney J. *Sociology 101* [course outline, Virginia Commonwealth University]. www.people.vcu.edu/~jmahoney/soc101sy.htm (accessed 15 November 2017).
5 Wills DD. *Basic ideas of McDonaldization* [course syllabus, Development Anthropology 2014, California Polytechnic State University]. www.cpp.edu/~ddwills/courses/ant352/352%20syllabus.html (accessed 18 December 2018).
6 PricewaterhouseCoopers. *The world in 2050*. www.pwc.com/gx/en/world-2050/assets/pwc-the-world-in-2050-full-report-feb-2017.pdf (accessed 16 January 2018).
7 PricewaterhouseCoopers. *Megatrends*. www.pwc.co.uk/issues/megatrends.html (accessed 15 January 2018).
8 Heinrich F. Corruption and inequality: How populists mislead people. *Transparency International*, 25 January 2017. www.transparency.org/news/feature/corruption_and_inequality_how_populists_mislead_people (accessed 16 October 2017).
9 Aldrick P. UK suffers its first lost decade in 150 years. *The Times*, 6 December 2016: 41.
10 Sylvester R, Thomson A. McMafia culture has spread to politics and the super-rich, says Misha Glenny. *The Times*, 6 January 2018: 36–37.
11 Transparency International. *Corruption perception index 2016*. www.transparency.org/news/feature/corruption_perceptions_index_2016 (accessed 9 January 2018).
12 Fullerton J. Xi jail for the corrupt runs out of space. *The Times*, 15 February 2018: 34.
13 Laing A. Ramaphosa pledges war on corruption. *The Times*, 17 February 2018: 40.
14 Kaufmann D, Kraay A, Mastruzzi M. *The worldwide governance indicators: Methodology and analytical issues* [World Bank Policy Research Working Paper 543, September 2010]. https://elibrary.worldbank.org/doi/abs/10.1596/1813-9450-5430 (accessed 20 July 2018).
15 World Bank Group. *Worldwide governance indicators*. www.worldbank.org/en/topic/governance/brief/anti-corruption (accessed 25 January 2018).
16 The World Bank. *Combatting corruption*. www.worldbank.org/en/topic/governance/brief/anti-corruption (accessed 19 December 2017).
17 Chartered Global Management Accountant and Transparency International. *Anti-corruption landscape 2017*. www.cgma.org/content/dam/cgma/resources/reports/downloadabledocuments/anti-corruption-landscape-2017.pdf (accessed 19 December 2017).

18 Pinker S. *Enlightenment Now: The Case for Reason, Science, Humanism, and Progress*. London: Penguin Books; 2018.
19 Samuel J. Our information age presents an existential threat to democracy. *Daily Telegraph*, 6 November 2017: 18.
20 Annan KA. How IT threatens democracy. *Project Syndicate*, 16 February 2018. www.project-syndicate.org/commentary/digital-threats-to-democracy-by-kofi-a-annan-2018-02 (accessed 19 February 2018).
21 Burleigh M. There's never been a worse moment in history for a parliament frozen in fear. The Mail on Sunday, 5 February 2017.
22 Huang Z. Your guide to understanding OBOR, China's new Silk Road plan. *Quartz*, 15 May 2017. https://qz.com/983460/obor-an-extremely-simple-guide-to-understanding-chinas-one-belt-one-road-forum-for-its-new-silk-road/ (accessed 12 December 2017).
23 The Economist reporter. What is China's belt and road initiative? *The Economist*, 15 May 2017. www.economist.com/blogs/economist-explains/2017/05/economist-explains-11 (accessed 5 June 2017).
24 Daley J. Peace with North Korea? Not very likely in a world full of death threats. *Sunday Telegraph*, 28 April 2018: 20.
25 Goldhill O. Humans are born irrational, and that has made us better decision-makers. *Quartz*, 4 March 2017. https://qz.com/922924/humans-werent-designed-to-be-rational-and-we-are-better-thinkers-for-it/ (accessed 15 March 2018).
26 Tversky A, Kahneman D. Availability: A heuristic for judging frequency and probability. *Cognitive Psychology* 1973; 5: 207–232. https://msu.edu/~ema/803/Ch11-JDM/2/TverskyKahneman73.pdf (accessed 28 October 2017).
27 PricewaterhouseCoopers. *From mall to mobile: Adjusting to new consumer habits* [Global consumer insights survey 2018]. www.pwc.com/gx/en/retail-consumer/assets/consumer-habits-global-consumer-insights-survey.pdf (accessed 28 April 2018).
28 Lueddeke GR, Kaufman GE, Kahn LH, Krecek RC, Willingham AL, Stroud CM, *et al*. Preparing society to create the world we need through 'One Health' education. *South Eastern European Journal of Public Health* 2016; VI: 1–4. www.seejph.com/index.php/seejph/article/view/122 (accessed 25 November 2017).
29 United Nations Development Programme. *Sustainable Development Goals*. www.undp.org/content/undp/en/home/sustainable-development-goals.html (accessed 20 October 2017).

8 Reorienting society towards a sustainable future

Future 'possibilities and dangers'

After 16 years in office (1975–1983 and 2011–2018), Governor Edmund Brown of California – with an economy ranked the sixth largest in the world and leaving the current budget with a US$ 5.4 billion surplus in a prospering state with the lowest unemployment rate (4.3 per cent) since 1976 – has good reason to be proud.[1] In his last 'State of the state address' at the end of January 2018, he reflected on progress in areas such as pensions, transportation, renewable energy, clean water, clean air, education, health care and higher education. He also recognised outstanding challenges in mental health, drug treatment programmes, the criminal justice system and education and training – all calling out 'for courage, for imagination and for generous dialogue'.[1]

Governor Brown's main message could be a wake-up call for California – a state that often accelerates change nationwide and, indeed, the world, exemplifying diversity of people, art and culture and innovations in technology. But global concerns are looming as Governor Brown compellingly acknowledged:

> These are truly exciting times, full of possibilities but also full of danger. Our world, our way of life, our system of governance – all are at immediate and genuine risk. Endless new weapon systems, growing antagonism among nations, the poison in our politics, climate change.[1]

Having lived through the consequences of drought, mudslides and fires on a truly sweeping scale, Governor Brown – backed by leading scientists at some of the most prestigious universities in the world – is convinced that climate change is the root cause of these natural disasters. His reminder that 'we should never forget our dependency on the natural environment and the fundamental challenges it presents to the way we live' is unwavering, as is his mantra: 'We can't fight nature. We have to learn how to get along with her'.[1]

Demonstrating that prosperity and caring for the environment are not at opposite ends of a dichotomy may possibly become Governor Brown's greatest achievement and political legacy. Another question that is equally pressing and critical that his successor (and other state leaders) may need to answer relates to how these two fundamental dimensions can be reconciled to ensure social coherence and stability.

The succinct reference in Governor Brown's address to 'possibilities' and 'danger' – with implications at state, national and global levels – provides urgency and relevance to achieving the UN-2030 Sustainable Development Goals (SDGs),[2] while continuing to work on achieving the previous Millennium Development Goals (MDGs)[3] and targets.

The United Nations global goals

The Millennium Development Goals (2000–2015)

Following the Millennium Summit of the United Nations in 2000, the adoption of the UN Millennium Declaration by 189 nations, including the eight MDGs, was hailed as a unique achievement in international development.[3] While raising the profile of global health, particularly in low- and middle-income countries, and making significant progress – extreme poverty was halved in 15 years; 43 million more children now go to school; new HIV infections decreased by 40 per cent and over 2 billion more people got access to clean drinking water – the MDGs 'were not achieved in their entirety'.[3]

Progress was uneven both between and within countries. With a population of over 1 billion people, Africa was and remains a case in point. Aside from children completing a full course in primary school and achieving gender equality in primary schools, none of the 12 main targets set for Sub-Saharan Africa were met. A key reason suggested for this lack of progress was that the MDGs fell far short in terms of addressing the broader concept of development encapsulated in the Millennium Declaration, which included human rights, equity, democracy, governance and, in particular, according to UN Secretary-General Ban Ki-moon, tackling root causes.[2,3]

The Sustainable Development Goals (2016–2030)

On 25 September 2015, 193 member states of the UN General Assembly ratified the UN-2030 Sustainable Development Goals (SDGs)[4] or UN-2030 Global Goals, as they are also called. As mentioned above, the 17 SDGs and 169 targets superseded the 2000–2015 UN MDGs.[2]

Extending the nature and scope of the MDGs dramatically, the SDGs (see Figure 8.1) are 'a universal call to action to end poverty, protect the planet and ensure that all people enjoy peace and prosperity'.[5] Framing the SDGs involved the largest consultation process within the history of the UN with contributions from more than a million people and an 'expert group' of over 3,000 participants from over 100 countries and 6 continents.

The 17 SDGs, which officially came into force in January 2016, are intended to be 'integrated and indivisible, global in nature and universally applicable' while 'respecting national policies and priorities'.[4] According to Johan

Figure 8.1 The UN-2030 Sustainable Development Goals.

Source: United Nations Development Programme (2015). www.undp.org/content/undp/en/home/sustainable-development-goals.html.

Rockström, Director of the Stockholm Resilience Centre, 'the SDGs are maybe the biggest decision in history … a much more complex agenda, which requires humans to reconnect with their planet.'[6]

The SDGs provide a synthesis of major global issues and place collaborative partnerships (SDG 17) at the centre of strategic implementation strategies.[7] Given the state of the planet, their adoption cannot come too soon. According to Marco Lambertini, Director General of World Wide Fund for Nature (WWF) International, transformative change is now imperative, challenging global leaders to respond to three main questions: 'What kind of future are we heading toward? … [W]hat kind of future do we want? Can we justify eroding our natural capital and allocating nature's resources so inequitably?'.[8]

Lambertini's concerns go beyond the immediate UN-2030 SDGs and demand finding, first and foremost, a lasting '*unity around a common cause* [italics added]'.[8] His message is intended for public, private and civil society sectors and implores these stakeholders to be proactive, to 'pull together in a bold and coordinated effort', for 'Heads of State' to think globally and for businesses and consumers 'to stop behaving as if we live in a limitless world' – before we face inevitable and potentially disastrous consequences.[8]

The 2015 Paris Climate Agreement is certainly testimony to Lambertini's urging being taken seriously and came on the heels of the groundbreaking 'Addis Ababa Action Agenda' at the UN Third International Conference on

Financing for Development in Ethiopia, which 'provided a financial foundation for implementing the global sustainable development agenda'.[9]

Indeed, the three historical milestones in 2015 – the Addis Ababa conference, the UN-2030 SDGs and the Paris Climate Agreement – represent major UN achievements, although translating the vision and goals into viable policies and enabling strategies, nationally and locally 'on the ground', presents considerable hurdles. Social, political and economic dichotomies remain and finding the 'middle ground', without sacrificing basic values and principles, for all stakeholders will be key to their success.

Report card on the SDGs – 2017

The 'Sustainable Development Goals Report 2017',[10] prepared by the UN Department of Economic and Social Affairs and launched at the same time as the Global SDG Indicators Database,[11,12] found that 'progress in meeting the SDGs' had 'not always been equitable' and that there were 'uneven advances across regions, between urban and rural populations and depending on gender, age and wealth'.[10]

Leila Mead, a consultant with the International Institute for Sustainable Development and the World Bank, prepared a synopsis of the report and concluded that 'increased efforts were needed to achieve the goals by 2030'.[13] Table 8.1 provides a summary of her review.

Table 8.1 Summary of SDG progress report 2017

SDG	*Progress made and remaining*
1 No poverty	Nearly one billion people have escaped extreme poverty since 1999, while three-quarters of a billion still remain in extreme poverty; working poverty has decreased, but is still pervasive in many regions and disproportionately affects youth; and less than half the world's population is covered by at least one social protection scheme, which vary by type of benefit.
2 Zero hunger	An extremely high number of children under the age of 5 suffer from malnutrition. Without accelerated progress, hunger will persist beyond 2030 and increased efforts are required to achieve the 2020 target on maintaining genetic diversity. The report further highlights lagging foreign and domestic official investments in agriculture.
3 Good health and well-being	Inter alia: child mortality has rapidly declined since 2000; the pace of progress in reducing maternal mortality must double to reach the 2030 target; only half of married women in Sub-Saharan Africa and Oceania satisfy their need for modern family planning; new infections of communicable diseases are declining due to targeted disease elimination and control programmes; and premature deaths from cancer, heart disease, diabetes and chronic respiratory disease are decreasing, but not fast enough to meet the 2030 target.

Table 8.1 Continued

SDG	*Progress made and remaining*
4 Quality education	Only 40 per cent of children participate in education one year before the start of primary school in poor countries; more than one-quarter of a billion school-aged children, adolescents and youth are not in school; poverty, gender and location of residence all influence reading proficiency; and lack of trained teachers and adequate facilities threaten education, especially in Sub-Saharan Africa.
5 Gender equality	Child marriage is declining, but remains common in Southern Asia and Sub-Saharan Africa; childbearing among adolescents in Sub-Saharan Africa remains high; female genital mutilation prevalence remains high in some countries, despite progress; and only half of women in selected countries make their own decisions regarding sexual relations, contraceptive use and health care.
6 Clean water and sanitation	The report calls for a new approach to measuring progress in drinking water, sanitation and hygiene; accelerated progress to end open defecation, especially in rural areas of Southern Asia and Sub-Saharan Africa; and more efficient ways of using water and producing food.
7 Affordable and clean energy	While 85.3 per cent of the global population now has access to electricity (up from 77.6 per cent) in 2000, more than one billion people, mostly in rural areas, still live without electricity. Other findings include: about three billion still lack access to clean and safe cooking fuels and technologies; and while renewable power generation is rising rapidly, comparable progress has not occurred in the heating or transport sectors.
8 Decent work and economic growth	Sustained high real economic growth remains elusive in least developed countries (LDCs); growth in labour productivity remains below its pre-financial crisis level; 1 in 10 children are still engaged in child labour although the situation is improving; and the expansion of ATMs (cash machines) provides access to financial services in many underserved regions.
9 Industry, innovation and infrastructure	The report describes slow manufacturing growth in LDCs; a decrease in emissions intensity of manufacturing; a growth in global investments in research and development (R&D); and almost universal mobile cellular coverage.
10 Reduced inequalities	Economic growth has helped reduce income inequality in many countries; voting rights of developing countries have increased in some international organisations, but are still below membership share; the trade community still grants more favourable access conditions to lower-income countries; and while sending remittances is still expensive, technology is leading to lower costs.

continued

Table 8.1 Continued

SDG	*Progress made and remaining*
11 **Sustainable cities and communities**	While the proportion of urban slum dwellers has decreased by 20 per cent since 2000, their actual numbers continue to grow; urban land expansion is outpacing population growth; municipal waste collection reaches only 65 per cent of urban dwellers; and more than 75 per cent of countries are engaged in coordinated urban planning.
12 **Sustainable consumption and production**	The report describes how in 2010, Australia and New Zealand had the highest 'material footprint' per capita, while Sub-Saharan Africa had the lowest, and that the material footprint per capita of developed regions far exceeds that of developing countries; most regions are using fewer resources per unit of production; and many countries are not meeting reporting commitments under chemical and hazardous waste environmental agreements.
13 **Climate action**	The report notes countries' commitments to taking climate action, implementing adaptation plans and enhancing resilience. The report further highlights that disaster risk reduction (DRR) can reduce deaths from small-scale events; and that disaster risks are exacerbated by poverty, inequality, environmental degradation and poor urban planning.
14 **Life below water**	Inter alia: marine protected areas (MPAs) have expanded; overfishing threatens about one- third of the world's fish stocks; and without concerted action, coastal eutrophication will increase in one-fifth of large marine ecosystems by 2050.
15 **Life on land**	The report explains that more biodiversity protection is occurring, although progress has recently slowed; forest area continues to shrink, but the rate of loss is slowing; corals, amphibians and cycads are facing the threat of extinction; and illicit wildlife poaching and trafficking are threatening the survival of many species.
16 **Peace, justice and strong institutions**	Inter alia: homicide rates have declined over the past decade, but are greater in countries with high income inequality; no region is immune from human trafficking; one-third of the world's prisoners are held without sentencing; companies in low-income countries experience more bribery requests than in wealthier countries; nearly one-third of the 102 journalists killed in 2016 were in Arab countries; more countries have adopted freedom of information legislation, but implementation is lagging; and more work is required to ensure that human rights institutions are compliant with international standards.
17 **Partnerships for the goals**	The report explains that while official development assistance (ODA) reached a new high, aid to the poorest countries is lagging; the debt service burden is increasing for low- and lower-middle-income countries; internet services remain inaccessible in much of the developing world; LDCs' share of global exports is declining in the merchandise trade but increasing in services; and only 17 countries, mostly in Europe and North America, have fully funded national statistical plans.

Source: International Institute for Sustainable Development. Equity and a *h*ealthy *p*lanet: IISD *a*nnual *r*eport 2016–2017. www.iisd.org/about/annual-reports.

Manufacturing industry response to the SDGs

In a white paper developed jointly by PwC, Global Manufacturing & Industrialisation Summit (GMIS) and United Nations Industrial Development Organization (UNIDO), the authors place responsibility for delivering the SDGs on both governments and business and explain 'why global manufacturing organisations must align their strategy with the SDGs'.[14] One of the first steps, they contend, is to recognise 'the relevance of the SDGs to their business activities', while the second encourages organisations to move 'away from "business-as-usual" in the way they deliver and create products and services'.[14] In particular, 'SDG goals 7, 8, 9, 12 and 13' may provide the 'greatest opportunities' and may be where 'their [manufacturers] biggest contribution' lies.[14] The paper also emphasises 'that the structural shift toward more innovation and technology-oriented industrial activities will significantly change the nature of competition, redefine work and redraw traditional industrial boundaries'.[14]

Among global manufacturing industry there is a growing consensus that 'Delivering the SDGs requires all countries to ensure that their policies are on track nationally and internationally' along with ensuring 'that the SDGs become a "whole-of-government" programme, not one confined to one or two ministries'.[14] In addition, according to country experiences, implementation of the SDGs requires 'reaching out to civil society and the private sector to consult them, even while the 2030 Agenda was still being negotiated for adoption by Heads of State and Government'.[14]

The PwC, GMIS and UNIDO white paper underscores commitment to the SDGs, referencing developments in the US, Germany, Rwanda, India and Columbia, in particular. Finland is presented as a case example in terms of implementation of Agenda 2030, building on two themes: 'the aim is a carbon-neutral, resource-wise and competent Finland where non-discrimination and equality are secured'.[14]

Of particular interest to other sectors, as summarised in Table 8.2, the white paper provides an analysis of the contribution of inclusive and sustainable industrialisation to each of the SDGs.

The response of governments

The 'Leaders' Declaration' at the G7 Summit in Germany between 7–8 June 2015, involving Japan, Canada, France, Germany, Italy, the UK, the US and the European Union, strongly affirmed high level commitment to 'the universal global sustainable development agenda'.[15] Group members committed to 'shared values and principles' and 'to uphold freedom, sovereignty and territorial integrity'.[15]

The 2016 G7 Summit in Japan focused on 'achieving economic empowerment for women, global health challenges and the cross-cutting theme of "leave no-one behind"'[16] and, according to Institute of Development Studies (IDS) Director, Professor Melissa Leach, underscored the importance of the G7 as 'key

Table 8.2 Manufacturing contributions to the SDGs

Sustainable Development Goal	*Industry's contribution*
1 **End poverty everywhere**	Manufacturing and productive activities crucially impact job creation, sustainable livelihoods, food security and equitable growth – key requirements for eradicating poverty.
2 **End hunger, improve nutrition and promote sustainable agriculture**	Support value addition in agro-industrial output, while generating job opportunities and increasing food security for rural communities.
3 **Attain healthy lives for all**	Support industrial small and medium-sized enterprises (SMEs) in the local manufacturing of essential drugs, medical supplies and optical products.
4 **Provide inclusive and equitable quality education and lifelong learning opportunities for all**	Development of entrepreneurial culture and skills, as well as young peoples' technical and learning capability, laying the groundwork for private sector development.
5 **Achieve gender equality and empower women and girls everywhere**	Creation of enabling environment for women entrepreneurs, and promotion of clusters and partnerships that facilitate access to information, technology and markets for women.
6 **Ensure availability and sustainable management of water and sanitation for all**	Transfer of best environmental practices to protect water resources and support the sound management of resource use in industrial activities.
7 **Ensure sustainable energy for all processes, along with enhanced utilisation of renewable energy sources**	Promotion of energy efficiency policies, technologies and production.
8 **Promote sustained, inclusive and sustainable economic growth, full and productive employment, decent work for all, diversification, export promotion and employment creation**	Building capacities of local industries for value addition and economic growth.
9 **Build resilient infrastructure, promote inclusive and sustainable industrialisation, and foster innovation**	Strengthening national capacities at the policy, institutional and enterprise levels in SME and entrepreneurship development, economic diversification, industrial upgrading, technological uptake and trade
10 **Reduce inequality within and among countries**	Support the provision of equitable and decent employment opportunities through substantial increases in industrial output, value-addition and transition to higher-tech production.

to realising the Sustainable Development Goals'.[17] The 2017 G7 Summit took place in Taormina, Italy, on 26 and 27 May. Group members discussed 'issues and crises that are most seriously threatening the security and well-being of our citizens and global stability, *inter alia,* foreign policy, the global economy, inequality, food security, and innovation and skills'.[18]

The 2018 G7 Summit was convened under the presidency of Canada in Charlevoix, Quebec, from 8–9 June 2018. The five themes discussed included: 'investing in growth that works for everyone; preparing for jobs of the future; advancing gender equality and women's empowerment; working together on climate change, oceans and clean energy; and building a more peaceful and secure world.'[19]

The summit, which was also attended by President Donald Tusk who along with President Jean-Claude Juncker represented the European Union, agreed on the following: to fight 'trade protectionism and to reform the World Trade Organization'; to urge 'Russia to end its actions undermining democracies and its support of the Syrian regime'; to call for 'the denuclearisation of the Korean peninsula and of Iran'; and to work together 'to achieve inclusive growth and gender equality'. In addition, 'the EU reaffirmed its strong commitment to implement the Paris climate agreement, along with France, Germany, Italy, Japan, and the UK.'[20]

While a final communique was adopted by the G7 at the end of the summit, it 'ended in acrimony' as 'the United States accused host Canada of stabbing President Donald Trump in the back'.[21] According to one report, what began as

> A fragile show of unity at the meeting in Quebec was blown apart with America isolated from the rest of the world's most industrialized democracies after Trump pulled support for a common communique in a showdown over his steel and aluminium tariffs.[21]

Generally-speaking, the G7 themes and associated communiques have been highly relevant to the progress of the SDGs. However, given the last meeting and other developments that seem to undermine the world order (e.g., nationalism, protectionism, alliances), the future of the G7, as presently constituted, remains uncertain. It may be noteworthy that the G8 became the G7 when Russia was suspended by the US and allies in March 2014 over the Ukraine crisis, more than a year before the UN 2030 SDGs were officially adopted by all 193 members of the United Nations General Assembly (UGA).

Concerns about global commitment to the SDGs were also raised at the G20 2016 and 2017 summits in China and Germany respectively. The G20 members (group of 19 major economies which includes G7 members) seem to have largely ignored key aspects of the 2030 Agenda – especially in areas related to ecology and human rights. The G20's main focus has primarily been on the economy, demonstrating a strong 'bias toward business'.[22]

In a summary report, Jens Martens, Executive Director of the Global Policy Forum (GPF), noted that many of the plans for action developed in preparation for the July 2017 G20 Summit – by the Employment Working Group, the Framework for Growth Working Group, the Health Working Group and nine others – 'played hardly any role in the political talks at the Summit itself and in media reports'.[22] The G20's main concerns were 'subordinated to the primacy of economic growth and the creation of investor-friendly framework conditions'.[22] The Executive Director's observation that the assumption 'what is good for the economy is good for society and the environment just won't work'[22] is likely shared by many.

A factor that might be preventing members of the G20, and perhaps other global and national decision-makers, from taking a more holistic (vs reductionist, overly rationalistic?) approach to setting global directions is the difficulty of seeing the interdependencies and interlinkages between the various elements that inform global issues or challenges. An example of this would be aligning the SDG and climate change agendas 'in terms of national implementation strategies, institutional coordination, financing, monitoring, evaluation and reporting'.[23] Making these connections is especially crucial in a world where silo thinking and planning continue to predominate and 'any little gain made in one could easily be eroded by the inaction in the other'.[23]

The three generally accepted dimensions of sustainable development – economic growth, social inclusion and environmental protection – make it clear that

> the current model of economic growth used by many countries, heavily based on the exploitation of natural resources, is not viable and there are many evidences which show that a more careful, that is, a more sustainable approach to the use of our limited resources, is needed.[24]

Indeed, the conventional wisdom on the relationship between the economy, people and the environment merits rethinking as it conveys the impression that economic growth (especially unregulated) drives prosperity, well-being and, in the final analysis, society. This dynamic may, however, be premised on essentially faulty logic, as Edward Lucas, a former editor at *The Economist* reminds us: 'For a start, we need to accept that business and finance are the servants of our civilization, not its masters'.[25]

A more accurate conceptualisation of planetary relationships is shown in Figure 8.2.[26] The model, developed by the Stockholm Resilience Centre, seems particularly relevant as, according to the authors, it

> changes our paradigm for development, moving away from the current sectorial approach where social, economic, and ecological development are seen as separate parts. *Now, we must transition toward a world logic where the economy serves society so that it evolves within the safe operating space of the planet* [italics added].[27]

While twenty-first century wealth derived largely from globalisation which has certainly been a major contributor to prosperity, there is considerable evidence (see Chapter 5) that it favours the few and is a major factor in inequality.[28] As a result, we are arguably seeing unforeseen political upheavals,

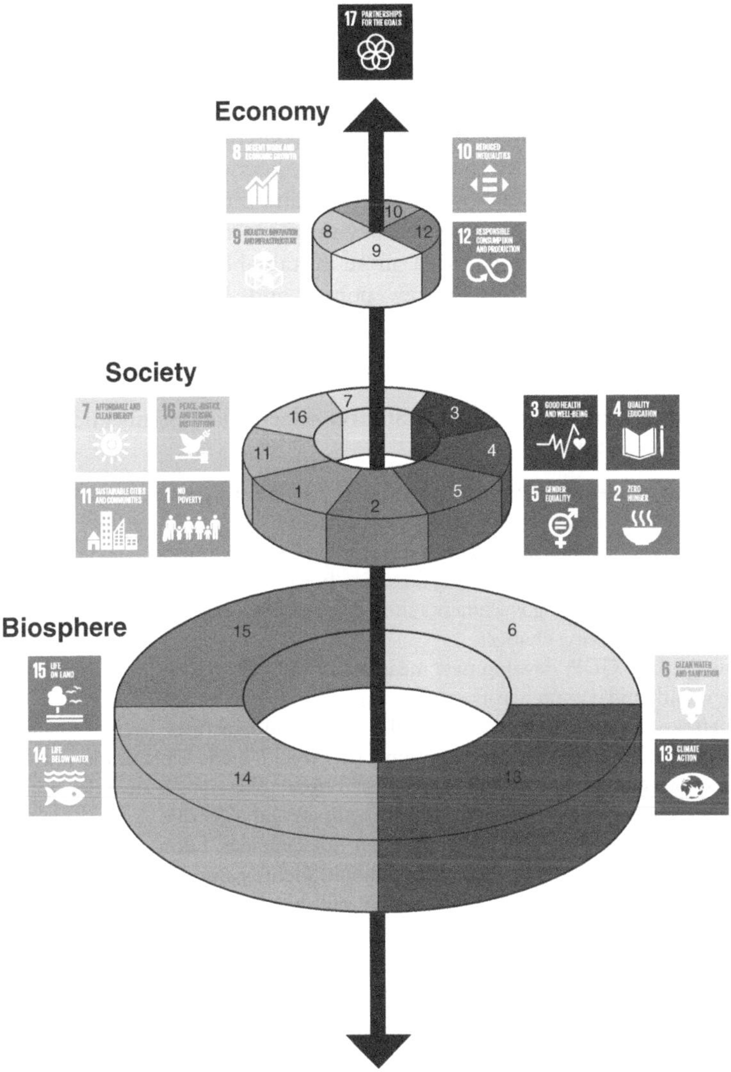

Figure 8.2 Paradigm shift – economy serving society.

Source: Azote Images (for Stockholm Resilience Centre).

borne to large extent out of frustration with or confusion about the status quo in the US, the UK, France, Germany, and possibly other nations that will added in the months and years ahead.

As discussed in Chapters 1 and 2, there is also growing anxiety that exploitation of the Earth's natural resources is the main cause of climate change and other associated consequences.

These concerns are not new. In 2012 the 'RIO+20 policy brief #6: 'Human well-being for a planet under pressure' highlighted that: 'Our rapidly increasing and urbanizing global population is facing unprecedented food, energy, economic and security crises, which are being compounded by climate change and extreme environmental events.'[28] The policy brief, produced by the scientific community to inform the UN Conference on Sustainable Development ('Rio+20'), argued that: 'As planetary boundaries are placed under increasing stress, so too are social bonds, relations and thresholds'.[28] It called for 'urgent, innovative solutions', setting out 'key messages and recommendations', to guide humanity on the road to a more sustainable socioeconomic and ecological future. Some of these key points and recommendations are summarised in Box 8.1.[28]

Box 8.1 Transition to social sustainability: selected key points and policy recommendations

- Both environmental and social sustainability should be regarded as imperative for overall human well-being by policymakers.
- For the Green Economy to succeed it needs to also focus on social and ecological sustainability as contributing factors to holistic well-being. International agencies and national governments should focus on growth in well-being and call for the following changes:
- move beyond GDP: develop new measures of progress, e.g. indices of inclusive wealth and comprehensive well-being;
- undertake a complete reform of global institutional structures: agencies such as the UN, Bretton Woods institutions, the World Trade Organization and others must be given new mandates that respect planetary boundaries and establish comprehensive human well-being as an end objective.
- Reducing absolute poverty is essential, but not sufficient. Efforts must also be taken to reduce inequality. Countries should identify key quantifiable well-being measures within the comprehensive well-being list, in which inequality is minimised through a participatory approach.
- Measures and incentives should be identified through a participatory, bottom-up process that recognises planetary boundaries and the social needs of all individuals.
- Economic systems should be developed as incentives for improving well-being, living within planetary boundaries and decreasing inequality.
- A trans-disciplinary research effort is needed to improve understanding of the links among comprehensive human well-being, ecological systems, socio-economic systems and pathways towards sustainability.

While the UN-2030 SDGs reflect the importance of the linkages between environmental and social sustainability, most of the remaining Rio+20 policy brief #6 recommendations have not been taken forward: for example, moving beyond GDP, developing new measures of progress and reforming 'the global institutional structure'. Political inertia or complacency could be part of the problem,[29] as free elections depend on voter support and their priorities – like their governments – are not necessarily in tune with global needs or threats (e.g., populism, protectionism), while autocratic leaders maximise their power and vested-interests and benefit most from destabilised societies.

Professor Gregory Treverton, former chair of the US National Intelligence Council, the agency that bridges the intelligence and policy communities, adds another perspective to these dynamics in a recent NIC report, 'Global trends: paradox of progress'.[30] Treverton points to the fallacy of 'the "prosperity presumption" that runs deep in most Americans but is often hardly recognized' and the assumption

> that with prosperity come all good things – people are happier, more democratic and less likely to go to war with one another. Yet, then we confront a group like ISIL [Islamic State of Iraq and the Levant], which shares none of the presumption.[30]

Clearly, as highlighted throughout this book, rethinking the sustainability of the planet is required as 'Changes in Earth Systems' and 'Human and Animal Health' are 'coming under increasing pressure', leading possibly to the most difficult question facing humanity today: '*How will political leaders and populations respond to a world less able to sustain life?*' [italics added].[30]

Alexander Likhotal, former president of Green Cross International and former advisor to the president of the USSR, Mikhail Gorbachev, holds that:

> In searching for new models for economic development, two important issues must be distinguished and addressed. First, how to produce more in order to meet increasing demand while making less of an impact on resources (often referred to simply as 'decoupling'). Second – even more fundamental – how to limit the increase in overall demand. The challenge is immense as currently we are in a completely contradictory situation where the more successful we are at promoting growth on the existing model, the greater and quicker will be the environmental and social disaster…
>
> When a system is fundamentally flawed, making it more efficient or accountable will not solve the problem. This model locks the world in continuing crisis, social injustice and the danger of environmental disaster. What we need today is to decouple economic growth from the use of energy and materials; simply increasing resource efficiency will not take us where we want to be. I am not questioning the objective of increasing energy and resource efficiency; essentially, we have no choice. What has to be questioned, however, is how production and consumption are being organized today.[31]

For Dr Likhotal one of the main barriers preventing the creation of new models of economic development relates to incorrectly measuring and valuing a wide range of goods and services essential to maintaining a safe, secure and sustainable planet. This systematic inadequate valuation (both under and over) is at the root of many major problems, such as the degradation of ecosystems, depletion of biodiversity and the destabilization of the social fabric of families and communities. Natural and social capital must be properly valued in economic terms in order for the economy to be 'real' and to be built on real values.[31]

Challenges facing the United Nations 2030 Sustainable Development Goals

As discussed earlier, the world's response to ongoing challenges is the UN-2030 SDGs agenda, whereby 'national political leaders' are able to 'generate strategic guidance calling for interdisciplinary relations across institutions' that is 'based on the concept of mutuality' and 'might induce states to find far greater unity in deliberations at the international level'.[4]

However, while the UN-2030 overall SDG vision for 'a world free of poverty, hunger, disease and want, where all life can thrive' is, as the UN planners admit, 'supremely ambitious and transformational',[32] it is also critical to the future of the planet but potentially difficult to achieve and raises fundamental questions,[31] such as:

- whether adequate resources will be available to enable coalitions of states and international organisations to address and lead programmatically on common challenges;
- whether the monitoring and compliance tools of international organisations will serve as confidence-building measures to reduce geopolitical tensions;
- the degree to which elites will be effective in guiding institutions and states through global transitions, and in promoting a strategic vision on crucial issues such as mitigating climate change and navigating global commons;
- the extent to which private actors will involve themselves in international rule making, enforcement, or dispute resolution – areas traditionally the responsibility of a state or public authority.[30]

In terms of the transformative change that is needed to achieve the SDGs and targets, many might also agree with the National Intelligence Council determination that

> There is no question that, political leaders will be key as only heads of state have the authority to press crossministerial agendas within their states. Such an approach will be a necessary counterbalance to what is now a very siloed international system.[30]

Along with political leaders, increasing importance is also being placed on the role of think tanks and academic institutions in achieving the SDGs. An

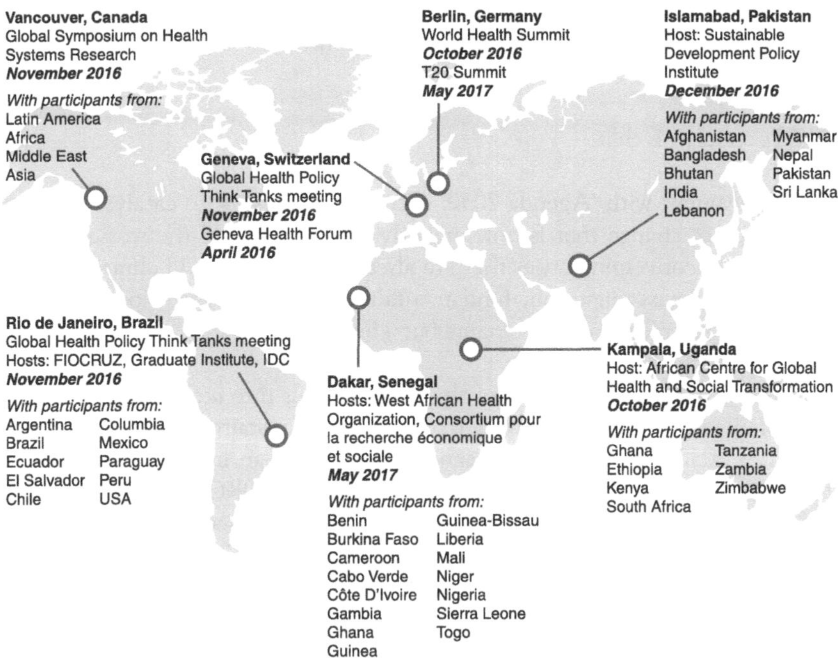

Figure 8.3 International Development Research Centre – regional consultations on achieving health for all.

Source: International Development Research Centre. *Four key findings to promote good health and well-being for all* (2017).

International Development Research Centre (IDRC) initiative is exploring health-related SDGs, which 'will influence the achievement of many of the other 16 SDGs, just as the progress toward these other goals will have a direct impact on health'.[33] In this regard, the IDRC carried out seven supported regional consultations in 2016 and 2017, as shown in Figure 8.3.

As summarised in Box 8.2, the IDRC identified four key findings[33] that could be extrapolated to the other SDGs.[32]

The SDGs and getting communities on board

In a timely paper, 'Making the Sustainable Development Goals work for local communities everywhere', Dr Daniel Wahl poses three questions that are fundamental to the success of the UN-2030 SDGs:

- Why are people not participating more enthusiastically in the widespread implementation of the Sustainable Development Goals?

- How can we engage communities everywhere to make the United Nations' Global Goals *their* local goals?
- How can individuals and collectives contribute to achieving 'Agenda 2030'?[34]

Wahl's concern is that

> engagement with 'Agenda 2030' is currently too slow to catalyze the magnitude of change that is now urgently needed to avoid the worsening of a series of converging crises that are already upon us – rapid climate change, famines, mass-migration, fundamentalism, resource wars, biodiversity loss and economic volatility to name but a few.[34]

He asserts that: 'We need to turn the Global Goals into *actively* shared objectives for all of humanity, if we want to create a more sustainable world that works for all by 2030.'[34] While 'leaders can – at best – set an intention on behalf of their electorate', the key to success – reinforced by IDRC findings – lies with 'effective collaboration between the public and private sector and civil society at local, regional and global scale' and with mobilising 'the active participation of people and communities around the world'.[34] Focusing on 'community participation in the implementation of the SDGs at sub-national levels', Wahl advocates the implementation of innovative learning initiatives such as 'Design for Sustainability' and 'Project Based Learning' programmes.[34]

Turning 'the top-down UN goals into meaningful projects that are locally relevant and can be collaboratively implemented by the communities themselves'[34] appears central to the success of the SDGs. This could, for example,

Box 8.2 International Development Research Centre – research findings on SDG implementation

- *Actions must be guided by accessible and quality data*: to understand SDG successes and failures at local, national and regional levels.
- *An intersectoral approach is key*: the 17 goals are interwoven and cannot be tackled in isolation … such as income, social environment, education, gender, and food security … only possible when different sectors collaborate around the common goal of improving well-being for all.
- *It is critical to engage civil society and citizens to solve health challenges*: emphasis should be placed on the importance of citizen engagement to address health challenges, especially those faced by the most marginalized. Citizens need a voice and the space to participate in decision-making processes to take charge of their health.
- *Robust and effective institutions are key players*: stakeholders such as policy research institutions … capable of combining rigorous research with contextual knowledge and citizen engagement to influence policy development.

also help to inform the IDRC's new Think Health Initiative,[33] whereby, together with the Graduate Institute, Geneva, 'policy research institutions are organizing regional working groups and conducting scoping studies on SDG implementation in South Asia, East Africa, Latin America, and West Africa.'[33]

Farooq Ullah, director and co-chair of UK Stakeholders for Sustainable Development, observes that it is necessary to replicate the motivational UN-2030 SDG 'esprit de corps at the regional, national and local levels where implementation truly takes form' as 'public policy alone will be insufficient to achieve the SDGs'.[35] In particular, he contends that attention needs to be paid to embodying, in practice, 'two key operational principles: those of Universality and Leaving No One Behind' as this is 'both unclear and challenging'.[35]

Moreover, among other 'elements that need to be addressed everywhere at the national and local levels', Ullah stresses that we must now focus on improving 'government and legislative machinery for sustainable development' and delivering 'formal and informal education, public awareness and capacity building training' as well as sharing 'learning between initiatives'.[35]

In particular, Ullah notes, 'In a time of seismic socio-economic shifts coupled with a growing sense of protectionism and tribalism', there is a need for 'a positive vision for the future that people can get behind'.[35] It is conceivable that the 'One Health and Well-Being' concept and approach might be an appropriate unifying thread that he, and perhaps others, are seeking.[33]

References

1 Brown EG. *Governor Brown delivers 2018 state of the state address: 'California is setting the pace for America'*. www.gov.ca.gov/2018/01/25/governor-brown-delivers-2018-state-of-the-state-address-california-is-setting-the-pace-for-america/ (accessed 10 January 2018).

2 United Nations. *The Millennium Development Goals report 2015*. www.un.org/millenniumgoals/2015_MDG_Report/pdf/MDG%202015%20rev%20(July%201).pdf (accessed 25 December 2015).

3 Lueddeke GR. Towards an integrative post-2015 sustainable development goal framework: Focusing on global justice – peace, security and basic human rights. *Southeastern European Journal of Public Health* 2014; II: 1–18. www.seejph.com/index.php/seejph/article/view/42 (accessed 16 February 2015).

4 United Nations (Sustainable Development Knowledge Platform). *Central global forum to advance action on Sustainable Development Goals set to highlight implementation progress and challenges*. www.un.org/sustainabledevelopment/blog/2017/07/central-global-forum-to-advance-action-on-sustainable-development-goals-set-to-highlight-implementation-progress-and-challenges/ (accessed 26 September 2017).

5 United Nations Development Programme. *The Sustainable Development Goals*. www.undp.org/content/undp/en/home/sustainable-development-goals.html (accessed 25 September 2017).

6 Lebada AM. *GEF event examines SDGs and planetary boundaries*. http://sdg.iisd.org/news/gef-event-examines-sdgs-and-planetary-boundaries/?rdr=climate-l.iisd.org (accessed 5 October 2015).

7 United Nations. *SDG 17*. www.undp.org/content/undp/en/home/sustainable-development-goals/goal-17-partnerships-for-the-goals.html (accessed 27 September 2017).

8 World Wildlife Fund. *Living planet report 2014 – Species and spaces, people and places*. http://assets.worldwildlife.org/publications/723/files/original/WWF-LPR2014-low_res.pdf?1413912230 (accessed 29 May 2016).
9 United Nations. *Addis Ababa action agenda*. https://sustainabledevelopment.un.org/frameworks/addisababaactionagenda (accessed 4 November 2015).
10 United Nations. *Sustainable Development Goals report 2017*. https://unstats.un.org/sdgs/files/report/2017/TheSustainableDevelopmentGoalsReport2017.pdf (accessed 29 October 2017).
11 United Nations. *SDG indicators*. https://unstats.un.org/sdgs/indicators/database/ (accessed 30 September 2018).
12 United Nations. *SDG indicators global database*. https://unstats.un.org/sdgs/indicators/database/ (accessed 4 December 2017).
13 Mead L. *SDG report 2017 provides overview of efforts to achieve SDGs – story highlights*. http://sdg.iisd.org/news/sdg-report-2017-provides-overview-of-efforts-to-achieve-sdgs/ (accessed 28 November 2017).
14 PricewaterhouseCoopers, United Global Manufacturing and Industrialisation Summit and United Nations Industrial Development Organization. *Delivering the Sustainable Development Goals: seizing the opportunity in global manufacturing*. White paper, 26 March 2017. www.pwc.com/m1/en/publications/delivering-sustainable-development-goals.html (accessed 12 February 2018).
15 German Federal Government. *Leaders' declaration – G7 summit, 7–8 June 2015*. www.g7germany.de/Content/EN/_Anlagen/G7/2015-06-08-g7-abschluss-eng_en.html (accessed 25 August 2016).
16 European Council of the European Union. *G7 summit, Ise-Shima, Japan, 26–27/05/2016*. www.consilium.europa.eu/en/meetings/international-summit/2016/05/26-27/ (accessed 10 September 2016).
17 Institute of Development Studies. *G7 are key to realising the Sustainable Development Goals*. www.ids.ac.uk/news/g7-are-key-to-realising-the-sustainable-development-goals (accessed 30 May 2016).
18 European Council of the European Union. *G7 summit in Taormina, Italy, 26–27/05/2017*. www.consilium.europa.eu/en/meetings/international-summit/2017/05/26-27/Goals (accessed 10 June 2017).
19 International Institute for Sustainable Development. *G7 leaders' summit, 8–9 June 2018*. http://sdg.iisd.org/events/2018-g7-leaders-summit (accessed 15 February 2018).
20 European Council of the European Union. *The Charlevoix G7 summit communique*. www.consilium.europa.eu/en/press/press-releases/2018/06/09/the-charlevoix-g7-summit-communique/ (accessed 9 June 2018).
21 Collinson, S. Trump plunges West into crisis over summit spa. *CNN*, 9 June 2018. https://edition.cnn.com/2018/06/09/politics/g7-donald-trump-foreign-policy/index.html (accessed 15 June 2018).
22 Martens J. *The G20 and the 2030 agenda: Contradictions and conflicts at the Hamburg summit*. www.globalpolicywatch.org/blog/2017/09/21/the-g20-and-the-2030-agenda/ (accessed 18 February 2018).
23 United Nations. *Linking climate change to the SDGs key to achieving the 2030 agenda and the Paris climate agreement*. https://sdgactioncampaign.org/2017/05/31/linking-climate-action-to-the-sdgs-key-to-achieving-the-2030-agenda-and-the-paris-agreement/ (accessed 15 June 2017).
24 Roberts J. *The Robert report*. World Sustainable Development Research and Transfer Centre. www.haw-hamburg.de/en/ftz-nk/about-us.html (accessed 27 February 2018).
25 Lucas R. Sacrifices are needed to keep civilisation afloat. *The Times*, 16 November 2015.
26 Rockström J, Sukhdev P. *How food connects all the SDGs*. www.stockholmresilience.org/ (accessed 29 December 2017).
27 Stockholm Resilience Centre. *A reference point for research on global sustainability*. www.stockholmresilience.org/ (accessed 22 January 2018).

28 International Council for Science. *Rio+20 policy brief #6: Human well-being for a planet under pressure – Transition to social sustainability*. www.icsu.org/cms/2017/09/Wellbeing_LR.pdf (accessed 20 October 2017).
29 Lueddeke GR. *Global Population Health and Well-Being in the 21st Century: Toward New Paradigms, Policy and Practice*. New York: Springer Publications; 2016.
30 National Intelligence Council. *Global trends – Paradox of progress*. www.dni.gov/index.php/newsroom/press-releases/item/1611-national-intelligence-council-releases-global-trends-report (accessed 9 January 2018).
31 Likhotal A. Environmental acceptability as the driver of new civilization. *CADMUS* 2014; 2(2): 24–34. www.cadmusjournal.org/node/389 (accessed 17 March 2015).
32 United Nations. *Transforming our world: the 2030 agenda for sustainable development*. https://sustainabledevelopment.un.org/post2015/transformingourworld (accessed 28 September 2015).
33 International Development Research Centre. *Four key findings to promote good health and well-being for all*. www.idrc.ca/en/article/four-key-findings-promote-good-health-and-well-being-all (accessed 14 November 2017).
34 Wahl DC. *Making the Sustainable Development Goals work for local communities everywhere*. https://medium.com/@designforsustainability/making-the-sustainable-development-goals-work-for-local-communities-everywhere-3f00bd5db31 (accessed 19 January 2018).
35 Ullah F. *Inclusiveness is the key to implementing the SDGs*. www.una.org.uk/5-inclusiveness-key-implementing-sdgs (accessed 6 December 2017).

9 *One Health and Well-Being* – inspiring a 'global unity of purpose'

Reimagining the health and well-being of society and the planet

The US National Intelligence Council (NIC) 'Global trends: paradox of progress' report makes for uncomfortable reading.[1] It projects, for example, that 'should current demographic, economic, and governance trends hold', we may be confronted with a 'more interconnected world' where 'differences over ideas and identities' divide rather than unite 'as the interplay between technology and culture accelerates and people seek meaning and security in the context of rapid and disorienting economic, social, and technological change'.[1] In short, for political leaders, appeals to identity coupled with technology will be 'useful for mobilizing supporters and consolidating political control'.[1]

In addition, the NIC portends that 'identity groups will become more influential. Growing access to information and communication tools will enable them to better organize and mobilize – around political issues, religion, values, economic interests, ethnicity, gender, and lifestyle'.[1] To exacerbate matters, it is anticipated that: 'The increasingly segregated information and media environment will harden identities – both through algorithms that provide customized searches and personally styled social media, as well as through deliberate … shaping efforts by organizations, governments, and thought leaders.'[1]

Th implications of these rising 'identity politics' scenarios are worrying. One of the most concerning may be 'the erosion of traditions of tolerance and diversity' across parts of the world, including 'the explicit use of nationalism and threatening characterizations of the West to shore up authoritarian control in China and Russia, and the inflaming of identity conflicts and communal tensions in Africa, the Middle East and South Asia'.[1]

Moreover, the NIC contends that governance may become even more difficult as 'Governments will increasingly struggle to meet public demands for security and prosperity' compounded by 'fiscal limits, political polarization, and weak administrative capacity'.[1] In addition, there are challenges from 'the changing information environment', public expectations and 'the proliferation of empowered actors who can block policy formation or implementation'.[1] The probable result, according to the NIC, will likely be 'growing public distrust and dissatisfaction. It will also increase the likelihood of protests, instability, and wider variations in governance'.[1]

'Identity politics' and divisions: are there constructive alternatives?

The US NIC projections appear to be based on several fundamental assumptions, including that:

- 'ideas and identities' must divide rather than unite (social);
- technology and culture must lead to political control rather than power-sharing (political);
- information and media environments must harden identities rather than act as vehicles for communication and understanding (technological);
- ecological priorities must continue to be subservient to economic progress (environmental and economic).

A common theme that runs through these premises is a view of the world as continuing to be 'a place made especially for humans and a place without limits' so that 'our task is to subdue and exploit the earth'.[2] But, *what if*, as Steven Pinker optimistically argues in *Enlightenment Now*, we could overturn the human 'pervasive negativity bias, driven by a morbid interest in what can go wrong' and 'defy the pessimists and improve the world'?[3]

And, *what if* – as discussed further in Chapter 12 on leadership – the world were to confront the actual forces (bullying tactics?) that are increasingly causing global deaths and anxieties[4] and may be responsible for undermining the future prospects of people and planet? *What if* those responsible were actually held to global account, and we were to replace current assumptions and feelings of resignation or apathy with alternative positive perspectives on, and actions for, sustaining the world and our role within it?

Pursuing new paths to the future through *One Health and Well-Being*

Demands for improving population health conditions in the early part of the twentieth century were largely responsible for the introduction of the concept and practices of public health.[4] This also brought about the separation of individual patient clinical care and public health, led to a large extent by Johns Hopkins University in the US. Since that time there has been a proliferation of 'healths', such as *public health*, *global health*, *planetary health*, *One Health*, thus 'bewildering' the public.[5]

To cut through some of the confusion, Dr Marija Cemma, a microbiologist at the Johns Hopkins Center for Health Security, interviewed 'global experts' to try to distinguish between these terms. Her investigation identifies three main camps: (1) those that deal with 'improving human health from the population perspective', transcending national borders, some including both preventive and individual-level clinical aspects (e.g. public health, global health); (2) those that are largely concerned with the 'sustainability of our civilisation' and 'resource consumption on the planet and human health' (e.g. planetary health, ecological health); and (3) those that encompass human, animal, plant and environmental health and well-being (e.g. One Health, One World–One Health).[5]

Of these, the insight that 'the health of people, other animals and the ecosystems of which we are a part are inextricably woven together'[6] goes back at least 2,500 years to ancient Greece and Hippocrates. Known as the 'father of medicine', Hippocrates 'urged physicians to consider where their patients lived, the foods they ate and waters they drank, their lifestyles, and the seasons of the year'.[4] According to the historical record, it then seems to have taken another few thousand years before Dr Rudolf Virchow (1821–1902), a German physician–pathologist (who coined the term 'zoonosis'), argued in the nineteenth century that 'there should be no dividing line between human and animal medicine, and that the causes of human disease were social and political'.[4] Other leading advocates of this approach[7] include:

- *Sir John McFadyean* (1853–1941), a UK veterinarian and physician considered the 'founder of modern veterinary research', who 'built bridges across human veterinary fields in infectious disease and comparative medicine'.[7]
- The US veterinarian *Dr. Calvin Schwabe* (1927–2006), considered the 'father of veterinary epidemiology', who framed the concept and coined the term 'One Medicine'.[8]
- *Dr. James Steele* (1913–2013), the 'father of veterinary public health' and founder of the US Centers for Disease Control and Prevention's (CDC) veterinary division in 1947. Steele co-authored *Confronting Zoonoses through Closer Collaboration between Medicine and Veterinary Medicine*[9] and advocated the One Health concept (referred to then as 'One Medicine') during much of the twentieth century, long before it became fashionable.

The main reasons for the delay in progressing the One Health (or an equivalent) concept in the twentieth century, at least, were made clear by David Waltner-Toews, a retired epidemiologist/veterinarian in the Department of Population Medicine at the University of Guelph, Ontario:

> Much of this integrative [human–animal] thinking was pushed into the background in the mid-twentieth century as many leaders and scholars were lured by the vision that infectious disease had been conquered, and that through basic scientific understanding, advanced technology and unlimited electrical power, humanity had somehow been freed from the bonds of nature. In the late 20th century, this vision was clearly demonstrated to be an illusion.[6]

The global wake-up call came in 1999 with the West Nile virus outbreak in New York City, which prompted the establishment of the National Center for Zoonotic [spread from animals to humans] Vector-Borne, and Enteric Diseases, now the CDC National Center for Emerging and Zoonotic Infectious Diseases (NCEZID).[10]

In the past few decades, One Health has been championed by a number of individuals,[7] such as:

- Veterinarian *Dr Roger Mahr*, former president of the American Veterinary Medical Association (AVMA) who, along with physician *Dr. Ronald Davis*, then president of the American Medical Association (AMA), passed a One Health resolution in 2007;
- *Dr Lonnie J King*, former director of the US CDC's new NCZVED,[10] past dean at the College of Veterinary Medicine and Professor of Preventive Medicine and current interim Vice President for Agriculture and Dean at the College of Food, Agricultural, and Environmental Sciences at Ohio State University;
- Veterinarian *Dr Bruce Kaplan* and medical researcher *Dr Laura Kahn*,[11,8] both of whom championed the establishment of the One Health Initiative Autonomous *pro bono* team (OHI);
- *Dr Thomas P Monath* – an internationally-recognised eminent physician, virologist and vaccinologist; identified as second among the 50 most influential people in the vaccines industry;[12]
- One Health 'trailblazers' *Dr Joann Lindenmayer*, chair of the One Health Commission (OHC) Board of Directors and *Dr Cheryl Stroud*, executive director, who along with dedicated colleagues have been spearheading global developments for the Commission,[13] and *Professor Ab Osterhaus*, chair of the One Health Platform (OHP).[14]

The individuals mentioned above, as well as many others, complement and support the deliberations and initiatives of such organisations as the World Veterinary Association (WVA), the World Medical Association (WMA), WHO, the World Organisation for Animal Health (OIE), the Food and Agriculture Organization of the UN (FAO), CDC, the Institute of Global Health – University of Geneva, the International Veterinary Students' Association (IVSA), the Network for the Evaluation of One Health (NEOH), the European Commission, One Health Central and Eastern Africa (OHCEA), the Southeast Asian One Health University Network (SEAOHUN), Chatham House and the World Bank Group, to name a number of leading organisations committed – along with many other supporters – to the *One Health* approach.

Dr Peter Rabinowitz, who directs the Center for One Health Research (COHR)[15] at the University of Washington, cites the Rockefeller Foundation report 'Safeguarding human health in the Anthropocene epoch', which 'outlines the case that anthropogenic changes in the environment are now threatening the basic life support services of the earth's systems', including 'biodiversity loss, climate change, particulate air pollution, ocean acidification, and deforestation'.[16] Rabinowitz highlights that these 'critical environmental threats' are 'intrinsic to our understanding and application of One Health' and stresses that 'the COHR approach to One Health is to look at the interconnections between hierarchically organized systems of human, animal, and environmental health'[17] (see Box 9.1).

Box 9.1 Planetary health is part of the *One Health* approach

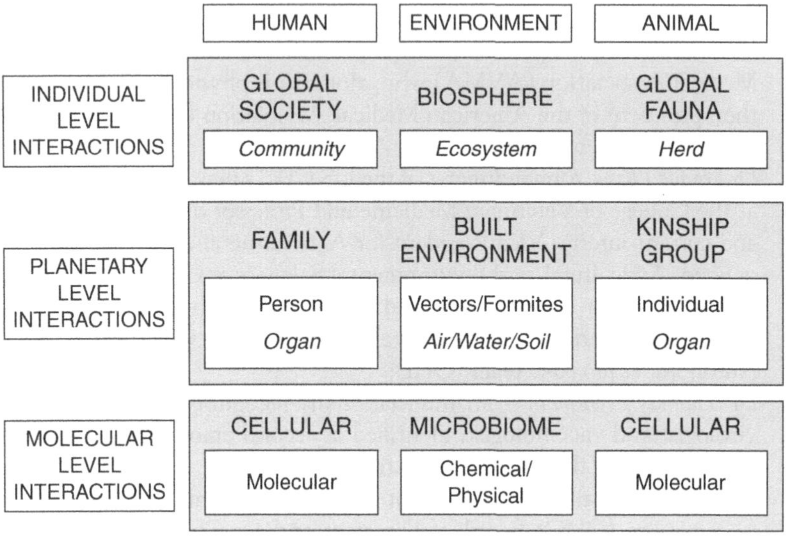

Adapted from Dr Peter Rabinowitz, Department of Environmental and Occupational Health Sciences, Center for One Health Research, University of Washington.

The World Bank *One Health* operational framework

When considering a global commitment to *One Health*, it is important to highlight the World Bank's recent comprehensive report, 'Operational framework for strengthening human, animal, and environmental public health systems at their interface'.[18] The report, 'developed with partners from EcoHealth Alliance' and which benefited 'from input from partner international organizations in the health, agriculture, environment and disaster risk reduction spheres', offers 'guidance to support countries in optimizing their One Health efforts and outcomes'.[18]

One of the report's many strengths is that it 'brings together tools and resources from different sectors (Figure 9.1) that can be leveraged for more comprehensive public health system capacity, including planning, implementation, and governance', thereby offering 'guidance to support countries in optimizing their One Health efforts and outcomes'.[19] Of particular importance in understanding the evolution of *One Health and Well-Being* is that the concept and approach now go beyond the zoonotic-medical paradigm (Figure 9.2) and are increasingly considered to be fundamental in addressing most other intractable sustainability challenges of the twenty-first century. Four 'take-aways' from the framework are summarised in Box 9.2.[19]

Box 9.2 Key 'take-aways' from the World Bank report 'Operational framework for strengthening human, animal, and environmental public health systems at their interface'

- **Starting points for One Health vary by context, disease and objectives.** Public health systems must be agile enough to address all hazards. While not all public health threats require a One Health approach, countries need strong human, animal and environmental health/management systems and coordination between them to determine which sectors are relevant for understanding and managing risk.
- **One Health is integral to the success of multi-sectoral national action plans for health security**, to address antimicrobial resistance and for disaster risk reduction. It can optimise pandemic preparedness planning and enhance climate change vulnerability assessments.
- **One Health approaches should be built into project design from the outset.** Engaging all relevant stakeholders early on can help optimise project success by promoting a common understanding of issues and joint solutions to address them, anticipating risks, targeting gaps, reducing duplication and facilitating relevant coordination channels.
- **Communication is a key priority for One Health understanding and implementation.** Planning for disease events and maintaining strong multi-sector coordination channels helps ensure consistent and effective messaging to manage risk, enhance efficiency and promote credibility of all sectors.

The authors of the World Bank report point to widening the scope of funding so that, 'In addition to infectious diseases, approaches outlined in the One Health Operational Framework can be applied to address other health hazards at the human–animal–environment interface, such as pollution, climate change, food insecurity, and more.'[19] They also note that 'to implement One Health approaches innovative financing options will need to be optimized' and that the World Bank 'supports countries in their efforts to prevent pandemics by strengthening veterinary and human health systems'.[19]

The report underscores that 'climate change is already impacting health' and 'that this damage is likely to continue, and that recent development gains are likely to be reversed without adequate investment in health sector resilience, with serious consequences for people and the environment'.[19] Exacerbating the impact of climate change, the authors warn that: 'Land degradation, another example of weak environmental health, can affect disaster resilience, safe water and food supplies, and pest control.'[19]

Reinforcing a recurring theme in this book, the World Bank report reminds us that: 'Strong coordination at the human–animal–environment interface can support multi-hazard preparedness to address existing threats and be flexible enough to anticipate threats on the horizon.'[19] As illustrated in Figure 9.1,[18] the report's concluding message that, 'By working across sectors to strengthen overall public health systems, we can reduce risks and help maximize healthy,

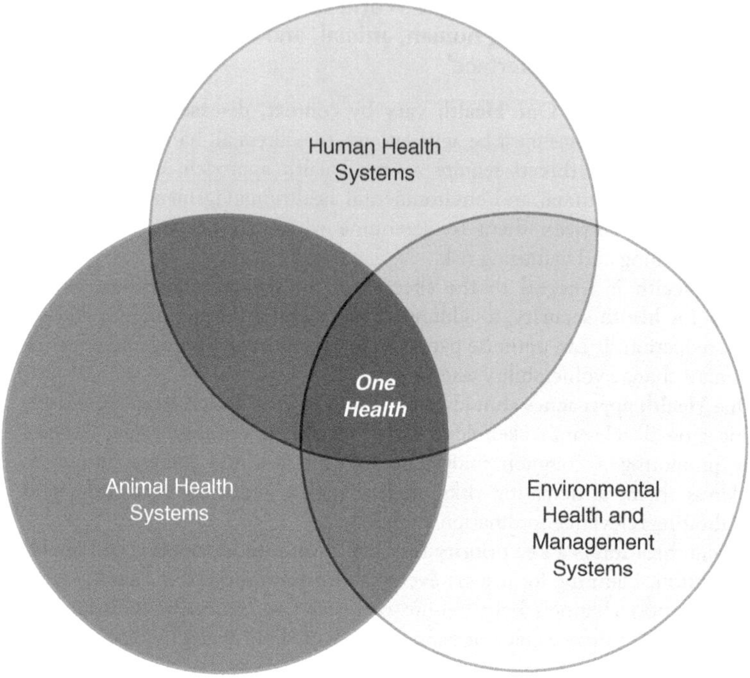

Figure 9.1 World Bank operational framework – disciplinary scope and areas of focus.
Source: World Bank Group. *Operational framework for strengthening human, animal, and environmental public health systems at their interface* (2018).

long-term and sustainable development',[19] is one that merits attention by all decision-makers, nationally and globally.

An appeal to world governing bodies

Overall, *One Health* is unquestionably an overarching concept that represents a means for achieving global 'unity around a common purpose'[20] and, as shown in Figure 9.2, by bringing together the health and well-being of people, animals, plants and the environment under one umbrella, bridging the human–animal interface supported by multi-disciplines and multi-sectors.

Inherent in the concept of One Health is the need to adopt ethical responsibilities that are rooted in interdependencies and the sanctity of life.[21] These were also at the heart of former UNESCO Director-General Irena Bokova's appeal, envisioned a few years ago, for 'a new humanism that reconciles the global and the local, and teaches us anew how to build the world'.[22]

As illustrated in Figures 9.3[23] and 9.4[24,25] moving in these directions – ensuring that the basic One Health values and principles of integration and

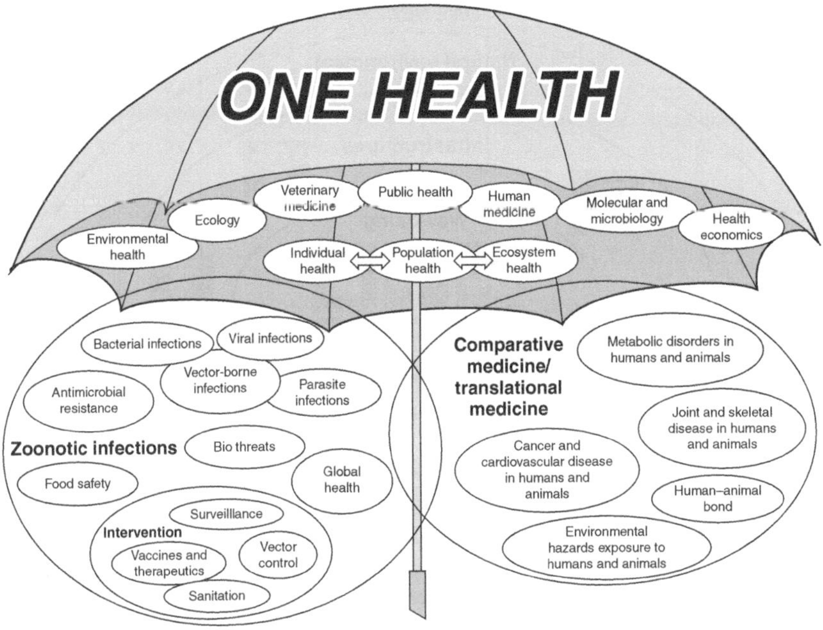

Figure 9.2 A *One Health* framework: zoonotic infections and comparative medicine/ translational medicine.

Source: One Health Initiative. www.onehealthinitiative.com/about.php.

collaboration guide legislation, policy and strategy in all fields of human endeavour – may give us the best chance yet to 'free the human race from the tyranny of poverty and want and to heal and secure our planet'.[26]

The interplay between institutions and belief systems

The late Nobel laureate, Douglass North, who applied economic theory to history, asked perhaps one of the most important and principled questions of our time: 'Why do some countries become rich while others remain poor?'.[27] Rather than following mainstream neoclassical economics, he 'emphasised the role of institutions – the rules and conventions of a society, such as laws, property rights, politics, customs and belief systems – in explaining differences in long-term economic growth' and 'integrated insights from different traditions in economics and different disciplines'.[27] Resonating with One Health principles of integration and collaboration, North's comparative static analyses 'drew on insights from psychology, politics and sociology to understand how human cognition, habits, religious beliefs, and ideology shape economic behaviour and, in turn, explains different long-term growth patterns'.[27]

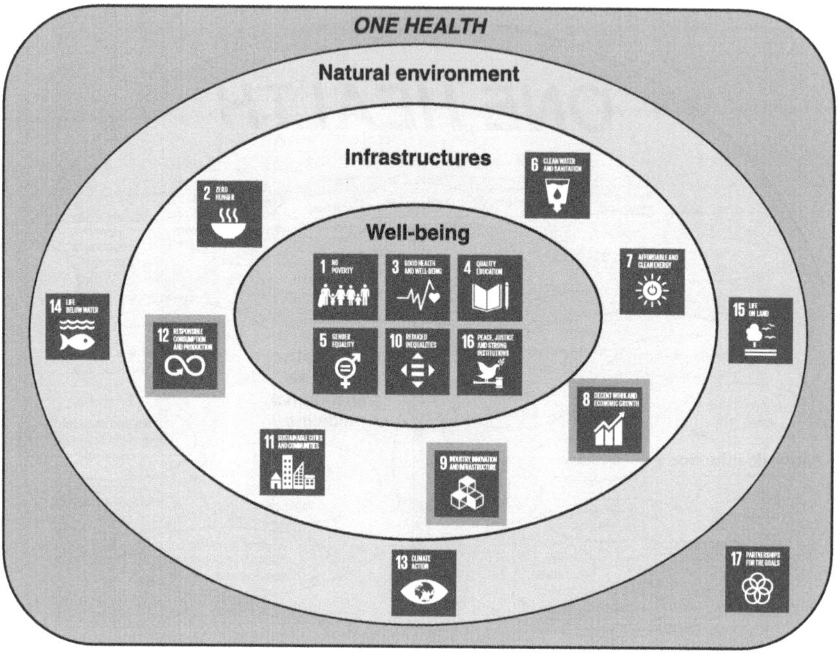

Figure 9.3 Framework grouping the Sustainable Development Goals (SDGs) based on their intended outcomes.

Source: adapted from Waage J, Yap C, Bell S, Levy C, Mace G, Pegram T, *et al.* Governing sustainable development goals: Interactions. In: Waage J, Yap C, editors. *Thinking Beyond Sectors for Sustainable Development*. London: Ubiquity Press; 2015.

In concluding his Nobel 1993 lecture, 'Economics through time',[28] North drew together the implications of 'development policy' for 'the creation of polities' – which may have relevance for implementation of the UN SDG agenda. He highlighted, among other things, that: 'Both institutions and belief systems must change for successful reform since it is the mental models of the actors that will shape choices.'[28] It is with these linkages in mind, together with North's reflections on the learning process as a means of 'understanding how particular belief systems emerge',[29] that the *One Health* concept and approach could make a distinctive contribution to the UN-2030 SDGs – in particular, North's observation that 'societies that get "stuck" embody belief systems and institutions that fail to confront and solve new problems of societal complexity'.[28]

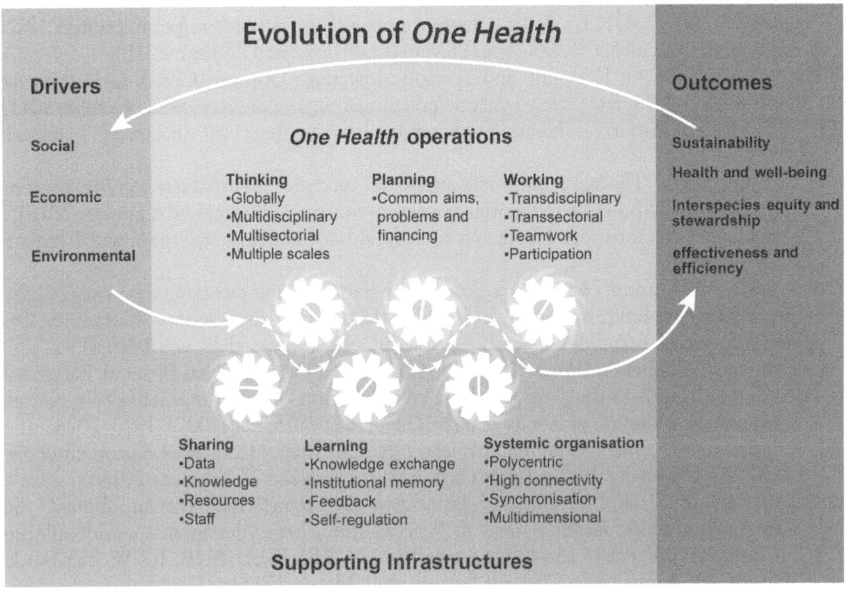

Figure 9.4 Relationship to *One Health* operations and infrastructure support.

Source: Rüegg SR, McMahon BJ, Häsler B, Esposito R, Nielsen LR, Speranza, CI, *et al.* A blueprint to evaluate one health. *Frontiers in Public Health* 2017; 5(20): 1–5.

References

1 National Intelligence Council. *Global trends – Paradox of progress*. www.dni.gov/index.php/newsroom/press-releases/item/1611-national-intelligence-council-releases-global-trends-report (accessed 9 January 2018).
2 Hanlon P, Carlisle S, Hannah M, Lyon A. *The Future Public Health*. Maidenhead, UK: McGraw Hill/Open University Press; 2012: 165.
3 Pinker S. *Enlightenment Now. The Case for Reason, Science, Humanism, and Progress*. New York: Viking; 2018.
4 Lueddeke GR. *Global Population Health and Well-Being in the 21st Century: Toward New Paradigms, Policy and Practice*. New York: Springer Publications; 2016: 12.
5 Cemma A. *What's the difference? A primer on evolving health disciplines*. www.globalhealthnow.org/2017-09/whats-difference-primer-evolving-health-disciplines (accessed 15 October 2017).
6 Waltner-Toews D. *One Health for one world: A compendium of case studies*. www.onehealthinitiative.com/publications/OHOW_Compendium_Case_Studies.pdf (accessed 5 December 2014).
7 Lueddeke GR. The UN-2030 Sustainable Development Goals and the *One Health* concept: A case for synergistic collaboration toward a common cause. *World Medical Journal* 2016; 62(4): 162–167. http://lab.arstubiedriba.lv/WMJ/vol.62/december-2016/#page=44 (accessed 19 July 2018).
8 Kaplan B. '*One medicine, one health*': *A brief overview*. www.onehealthinitiative.com/publications/Sarasota%20Friendship%20Center%20Speech%20PP%20Slides%20Bruce%20Kaplan%20%20Jan%2011%202008.pdf. (accessed 15 March 2015).

9 Kahn LH, Kaplan B and Steele JH. Confronting zoonosis through closer collaboration between medicine and veterinarian medicine (as 'one medicine'). *Veterinaria Italiana* 2007; 43(1): 5–19. http://onehealthinitiative.com/publications/01%20Kahn%20et%20al%20%205-19%208Mar07.pdf (accessed 28 June 2018).
10 National Center for Emerging and Zoonotic Infectious Diseases (CDC). *Our work, our stories – 2011–2012.* www.cdc.gov/ncezid/pdf/annual-report.pdf (accessed 25 October 2013).
11 One Health Initiative. *Home.* www.onehealthinitiative.com/ (accessed 7 January 2018).
12 VaccineNation. *The most influential people in vaccines.* http://vaxess.com/wp-content/uploads/2016/01/the-most-influential-people-in-vaccines.pdf (accessed 8 January 2018).
13 One Health Commission. *Home.* www.onehealthcommission.org/ (accessed 7 January 2018).
14 One Health Platform. *Home.* https://onehealthplatform.com/ (accessed 8 January 2018).
15 Department of Environmental and Occupational Health Sciences. *Center for One Health Research.* http://deohs.washington.edu/cohr/ (accessed 20 April 2018).
16 Whitmee S, Haines A, Beyrer C, Boltz F, Capon AG, De Souza Dias BF, et al. Safeguarding human health in the Anthropocene epoch: Report of The Rockefeller Foundation–Lancet Commission on planetary health. The Lancet 2015; 386(1007): 1973–2028.
17 Rabinowitz P. *One health and planetary health.* http://deohs.washington.edu/cohr/blog/2017/15/one-health-and-planetary-health (accessed 26 February 2018).
18 World Bank Group. *Operational framework for strengthening human, animal, and environmental public health systems at their interface.* http://documents.worldbank.org/curated/en/703711517234402168/pdf/123023-REVISED-PUBLIC-World-Bank-One-Health-Framework-2018.pdf (accessed 11 March 2018).
19 Voegele J, Evans T, Kemper KE. *One health approach is critical to de-risk human, animal and environmental health.* http://blogs.worldbank.org/health/one-health-approach-critical-de-risk-human-animal-and-environmental-health (accessed 12 March 2018).
20 Lueddeke GR, Kaufman GE, Lindenmayer JM, Stroud CM. Preparing society to create the world we need through 'One Health' education – Report of a global survey and web conference on 'One Health' K-12 education, 18 November 2016. *South Eastern European Journal of Public Health* 2017; VIII: 1–22. www.seejph.com/index.php/seejph/article/view/142 (accessed 19 January 2018).
21 UNESCO. Participation at the 35th UNESCO general conference [speech by Director-General Irena Bokova]. http://natcomreport.com/Tanzania/pdf-new/participation.pdf (accessed 17 May 2015).
22 Earth Charter International. *The earth charter.* http://earthcharter.org/discover/the-earth-charter/ (accessed 17 January 2018).
23 Waage J, Yap C, Bell S, Levy C, Mace G, Pegram T, *et al.* Governing sustainable development goals: Interactions. In: Waage J, Yap C, editors. *Thinking Beyond Sectors for Sustainable Development.* London: Ubiquity Press; 2015: 79–88.
24 Queenan K, Garnier J, Nielsen RL, Buttigieg S, de Meneghi D, Holmberg M, *et al. Roadmap to a one health agenda 2030.* www.cabi.org/CABReviews/FullTextPDF/2017/20173134856.pdf (accessed 25 November 2017).
25 Rüegg SR, McMahon BJ, Häsler B, Esposito R, Nielsen LR, Speranza, CI, *et al.* A blueprint to evaluate one health. *Frontiers in Public Health* 2017; 5(20): 1–5.
26 United Nations. *Transforming our world: the 2030 agenda for sustainable development.* https://sustainabledevelopment.un.org/post2015/transformingourworld (accessed 28 September 2015).
27 Nielsen K. Douglass North obituary. *Guardian*, 27 January 2016.
28 North DC. Economic performance through time. *The American Economic Review* 1994; 84(3): 359–368.
29 Faundez J. *Douglass North's theory of institutions: Lessons for law and development.* http://wrap.warwick.ac.uk/84086/1/WRAP_%253A10.1007%252Fs40803-016-0028-8.pdf (accessed 25 October 2017).

Part III

Creating a 'more just, sustainable and peaceful world'

10 Global education

Building foundations for the future

UN global goals – well-being and education

It is almost impossible to think about SDG 16 – the promotion of 'just, peaceful and inclusive societies',[1] which underlies the title of Part III of this book – without also reflecting on SDG 3 and SDG 4. The former is concerned with ensuring 'healthy lives' and promoting the 'well-being for all at all ages'[2] while SDG 4 is about 'quality education' as 'a fundamental human right' and 'promoting lifelong learning'.[3] Both are of utmost importance to 'building foundations for the future' and, as the UN-2030 Global Goals pledge, 'leaving no one behind'.[4]

Although SDG 3's concern with health and well-being is extremely relevant to this chapter's main theme, given space constraints, the chapter will focus mainly on SDG 4: education. Both are at risk of being undermined in view of barriers – mainly, but not exclusively, in developing nations – such as inequality, poverty, finance, extremism and continuing regional armed conflicts, while the more advanced economic nations are facing the more immediate challenge of preparing students for a world that is undergoing rapid fundamental social, economic and technological change.

Structurally, Part III addresses approaches for achieving a sustainable future through a combination of education – including increasing awareness of the world we inhabit – visionary leadership and optimising the resourcefulness and commitment of the younger generation to enable the continuation of 'a core of ideas, practices and institutions' to 'provide a bedrock'[5] for our civilisation while ensuring the well-being of our planet. To these ends, this chapter centres on two key educational dimensions: *The UN Education Agenda – 2030* and *Preparing Learners for the Twenty-First Century*.

The UN Education Agenda – 2030

Children and youth denied the right to an education worldwide

Although the UN 2015 'Millennium Development Goals Report' highlighted 'notable advances in primary school net enrollment rates and decreases in the number of out-of-school students', it also noted that 'getting an education – let alone a quality education – remains far out of reach'.[6]

As shown in Figure 10.1,[7,8] it is estimated that

> 263 million children and youth are out of school ... equivalent to a quarter of the population of Europe. The total includes 61 million children of primary school age, 62 million of lower secondary school age, and includes the first ever estimate of those of upper secondary school age at 141 million [about twice the population of Britain].[9]

UNESCO reports that

> Of all the regions, sub-Saharan Africa has the highest rates of exclusion. Over a fifth of children between the ages of 6–11 are out of school, followed by a third of youth between the ages of 12–14 ... almost 60% of youth between the ages of 15–17 are not in school.[9]

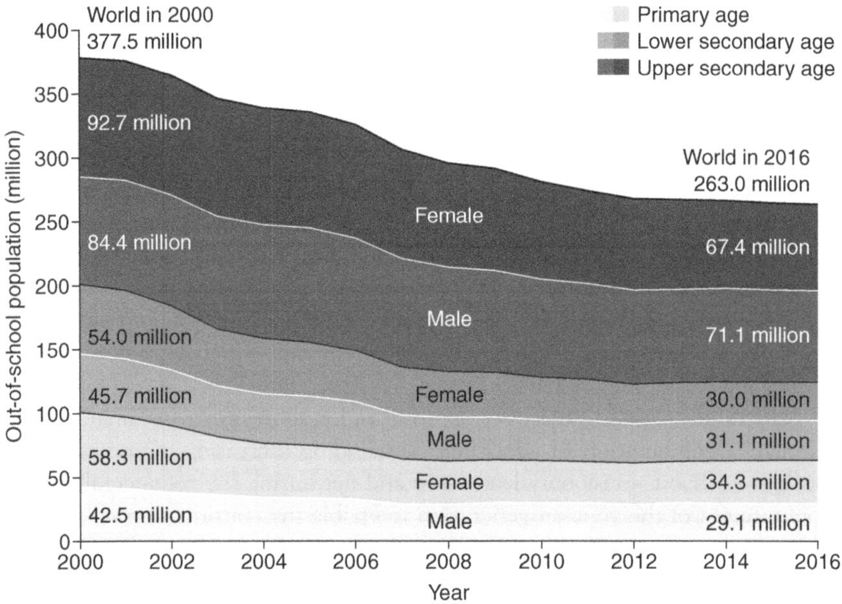

Figure 10.1 Global number of out-of-school children, adolescents and youth, 2000–2016.

Source: UNESCO Institute for Statistics database. http://uis.unesco.org/sites/default/files/documents/fs48-one-five-children-adolescents-youth-out-school-2018-en.pdf.

Note

'In 2016, 263 million children, adolescents and youth were out of school, representing nearly one-fifth of the global population of this age group.' However, while those excluded from education decreased between 2000 and 2010, this decrease 'essentially stopped in recent years', declining only 'by little more than 1 million per year since 2012' (UNESCO Institute for Statistics).

In addition, it is known that: 'In every region, older youth face greater barriers to education. According to the global average, 15 to 17-year-olds are four times more likely to not be in school than children between the ages of 6–11.'[9] UNESCO Institute for Statistics (UIS) data also confirms that 'girls are more likely than boys never to set foot in a classroom, despite all the efforts and progress made over the past two decades'.[9] Moreover, '15 million girls of primary-school age will never get the chance to learn to read or write in primary school compared to about 10 million boys. Over half of these girls – 9 million – live in sub-Saharan Africa.'[9]

Exacerbating the current situation, as discussed in Chapter 6, are the continuing armed conflicts in many regions of the world. As a result, it is estimated that: 'Globally, 35% of all out-of-school children of primary age (22 million), 25% of all out-of-school adolescents of lower secondary age (15 million), and 18% of all out-of-school youth of upper secondary age live in conflict-affected areas (26 million).'[9] Furthermore, more than '5.3 million people, including 2.5 million children, have been living as registered refugees in Turkey, Lebanon, Jordan, Iraq and Egypt'.[10]

The Incheon Declaration for Education 2030

The World Education Forum, hosted by the Republic of Korea in May 2015, was supported by UNESCO together with UNICEF, the World Bank, the United Nations Population Fund (UNFPA), UNDP, UN Women and the United Nations Refugee Agency (UNHCR). The conference, attended by over 1,800 participants from 160 countries, resulted in the 'Incheon Declaration for Education 2030'[11] – which set out a new vision for education to 2030 (see Box 10.1). Recognising the potential global impact of the conference, attendees also included 'over 120 Ministers, heads and members of delegations, heads of agencies and officials of multilateral and bilateral organizations, and representatives of civil society, the teaching profession, youth and the private sector'.[11]

Box 10.1 The Incheon Declaration for Education 2030[11]

> 'Our vision is to transform lives through education, recognizing the important role of education as a main driver of development and in achieving the other proposed SDGs. We commit with a sense of urgency to a single, renewed education agenda that is holistic, ambitious and aspirational, leaving no one behind. This new vision is fully captured by the proposed SDG 4 "Ensure inclusive and equitable quality education and promote lifelong learning opportunities for all" and its corresponding targets ... *It is inspired by a humanistic vision of education and development based on human rights and dignity; social justice; inclusion; protection; cultural, linguistic and ethnic diversity; and shared responsibility and accountability*' [emphasis added].

Given that education is a basic human right, the adoption of 'the Incheon Declaration represents the firm commitment of countries and the global education community to a single, renewed education agenda – Education 2030 – which is holistic, ambitious, inclusive and aspirational'.[11]

Guiding the implementation of the SDG 4 agenda is the 'Education 2030 Framework for Action' (FFA),[12] which 'was developed through a highly consultative process with its essential elements agreed upon at the World Education Forum held in Incheon, Republic of Korea' and 'formally adopted and launched at a high-level meeting, alongside the 38th UNESCO General Conference at UNESCO headquarters, Paris on 4 November 2015'.[12]

The FFA aims at

> mobilizing all countries and partners around the Sustainable Development Goal (SDG) on education and its targets, and proposes ways of implementing, coordinating, financing and monitoring SDG4-Education 2030 to ensure inclusive and equitable quality education and lifelong learning opportunities for all.[12]

The 2017/2018 Global Education Monitoring Report

The most recent UNESCO 'Global Education Monitoring Report' (hereafter '2017/8 GEM Report')[13] was issued in July 2017 and focuses on 'the role of accountability in global education systems regarding achievement of the vision of UN Sustainable Development Goals'.[4]

Framed within the SDG 4 targets, the '2017/8 GEM Report' ranges from reviews of 'how all actors in education are currently held to account'[13] to monitoring of progress in SDG 4 to recommendations for taking the '2017/8 GEM Report' forward from 2018/2019 towards 2030. As selected indicators below illustrate, significant challenges remain across all targets but targets 4.1, 4.3, 4.7, 4a and 4c appear to be the most pressing.

Only 40 per cent of children participate in education one year before the start of primary school in poor countries. Moreover, as mentioned earlier, more than 264 million school-aged children, adolescents and youth are not in school; social determinants, poverty, gender and location of residence are all factors in reading proficiency; and lack of trained teachers and adequate facilities undermine the quality of education – especially in Sub-Saharan Africa nations.

Essential skills, higher education, employment and social mobility

There can be little doubt that a lack of basic skills is a critical factor in preventing many young people from entering advanced or higher education. For example, only 8 per cent of young people in Sub-Saharan Africa are enrolled in higher education compared with 75 per cent in Europe and Northern America.[13] As things stand, this low participation level could severely compromise future economic growth, prosperity and social well-being.

As previously reported, a 'major problem for Africans' – 'a dramatic disconnect' – is between 'the continent's youth', those 'who yearn for change and greater life chances', and 'Africa's longest serving leaders'.[14] This dissonance is exacerbated by the fact that '70% of African citizens [are] younger than 30 years' and demand not only 'democratic representation and economic participation' but also 'greater access to education and to skills that will lead to employment'.[15]

In this regard the authors of 'Waiting for happiness in Africa'

> raise a crucial question that has major implications for Europe as well as the continent: what will happen to Africa's youth who do not find jobs in their countries of birth by their mid to late twenties? Will they despair, join extremist groups, or emigrate?[15]

Many have decided to leave the continent – over '80,000 of the passengers on the Mediterranean people-smuggling boats have come from sub-Saharan Africa' – to seek a better life, thereby, as the authors note, overwhelming Europe.[15]

Moreover, 'There are over 200 million young people aged 15 to 24 at present, constituting 37% of Africa's labour force but (60% of the continent's total).'[15] Thus, there is little doubt that 'Africa will need to provide jobs for its youth if it is to meet their aspirations for the good life'.[15] It is estimated that '18 million jobs will need to be created every year just to accommodate Africa's current jobseekers'.[15]

According to the Programme for Infra-structure Development in Africa, the development of the continent's infrastructure with investments of 'up to $93 billion a year until 2020',[15] could be a key catalyst for providing employment opportunities. Unquestionably, major investment is needed in Africa's young people to address barriers to youth employment and 'avert a population time bomb'.[15] To their credit, the authors of 'Waiting for happiness in Africa' conclude on a positive note, emphasising that Africa is an optimistic continent and that this hopefulness is shared 'most of all by the youth who have their lives ahead of them … [and] might serve as a self-fulfilling prophecy for the continent'.[15]

Dr Jordan Naidoo, director of UNESCO Education, claims there is a consensus that achieving the SDG 4 targets 'requires country-level action linked to existing national (only if necessary) new contextually defined plans based on an assessment of the current situation'.[16] Moreover, he observes that 'the challenge of addressing teacher shortages or quality has not been fully comprehended'.[16] Key findings from 'The Sustainable Development Goals Report 2017' are summarised in Table 10.1.[17]

Table 10.1 SDG 4: quality education targets and 2017 progress update

Targets	*Progress*
4.1 **Primary and secondary education** (proficiency in reading and mathematics)	'Fewer than 50 percent of students complete secondary school … [this] unquestionably has a negative effect on participation in tertiary education, which has increased by 30 percentage points in upper middle income countries from 17% to 46%, [but] has stagnated in the Caucasus and Central Asia and in sub-Saharan Africa.'
4.2 **Early childhood**	Only 'one-third of countries worldwide legally stipulate[d] at least one year of free provision, 21% one year of compulsory provision and 17% one year both free and compulsory' … in 2015'.
4.3 **Technical, vocational, tertiary and adult education**	'Most regions have seen little change in enrolment in technical and vocational education since 2000, although participation rose in the Caucasus and Central Asia and fell in the Pacific. Technical and vocational education remained male-dominated, with girls accounting for 43% of enrolment.'
4.4 **Skills for work**	'According to data on the global indicator, adults had not performed even basic computer tasks in low and middle income countries.'
4.5 **Equity**	'On average, the world has achieved the target of gender parity at all levels except tertiary education. However, this is not true of all regions and country income groups or at the individual country level. Only 66% of countries have achieved gender parity in primary education, 45% in lower secondary education and 25% in upper secondary education.'
4.6 **Literacy and numeracy**	'Between 2000 and 2015, the adult literacy rate increased from 81.5% to 86%, although it remains at 64% in sub-Saharan Africa and just below 60% in low income countries. The number of adults with no literacy skills has fallen by just 4% to 753 million.'
4.7 **Sustainable development and global citizenship**	'While peace/non-violence, human rights/fundamental freedoms and cultural diversity/tolerance fared better in the 2012 study, education for sustainable development has received very little attention to date.'
4.a **Education facilities and learning environments**	'In sub-Saharan Africa, only 22% of primary schools have electricity; access to drinking water was below 75% in 72 of 148 countries. Access to basic sanitation facilities was below 50% in 24 of 137 countries, including 17 in sub-Saharan Africa.'
4b **Scholarships**	'Among those that do [support scholarships], spending decreased from US$1.2 billion in 2010 to US$1.15 billion in 2015, or by 4%, on par with the overall decrease in aid to education. Australia (23%), France (14%) and the EU institutions (15%) account for over half of all scholarship aid. In addition, donors report imputed student costs, which reached US$1.76 billion in 2015.'
4c **Teachers**	'Globally, 86% of primary school teachers are trained; the proportion is lower in Southern Asia (77%), the Caribbean (70%) and sub-Saharan Africa (62%).'

Source: adapted from United Nations, *Sustainable Development Goals Report 2017*. https://unstats.un.org/sdgs/report/2017.

Special challenges with SDG Education Target 4.7: 'Sustainable development and global citizenship'[18]

Only a few countries have integrated learning about sustainable development in terms of policy, curricula or teacher education (7 per cent!). There appear to be three fundamental issues:

1. 'There is still no consensus on what outcomes education for sustainable development and global citizenship education should achieve.'[13]
2. The themes focus only on human interactions, not relationships with the planet and other species.
3. Given the state of the world, we need to question the extent to which present approaches to teaching these topics are transferring from classrooms to real life, or whether they are making any difference to the health and well-being of people and the planet.

Progressing the '2017/8 GEM report'

The '2017/8 GEM report' emphasises, among other things, that: 'Accountability in education starts with governments, which bear the primary duty to ensure the right to education.'[13] 'Building trust and a shared understanding of respective responsibilities with all education actors' and developing 'credible education sector plans and transparent budgets with clear lines of responsibility and truly independent auditing mechanisms' are priority accountability mechanisms.[13]

Making transparent, relevant and timely information available, providing financial resources and building capacity – skills and training[18] – are key in implementing the '2017/8 GEM report'.

Funding education

On 20 September 2017 at the high-level event 'Financing the future: Education 2030'[19] that took place during the seventy-second session of the UN General Assembly in New York, UN Secretary-General António Guterres, UN Deputy Secretary-General Amina Mohammed, the heads of state of France, Norway, Malawi and Senegal as well as other global education leaders committed to tackling the global 'education crisis'.

Attending the high-level 'Financing the future' meeting, Malala Yousafzai, Malala Fund co-founder and 'UN Messenger of Peace', highlighted that

> 130 million girls are out of school today. They are pushing back against poverty, war and child marriage to go to school. The Sustainable Development Goals were a promise that we would fight with these girls. So far, we have failed. We have big goals, but we will not reach any of them unless we educate girls. If we want to grow economies, improve the air we breathe, promote peace and advance public health, we must invest in girls.[20]

Resonating with the arguments of attendees for increased financial support, Secretary-General Guterres emphasised that: 'Investing in education is the most cost effective way to drive economic development, improve skills and opportunities for young women and men, and unlock progress on all 17 Sustainable Development Goals.'[20] Former UK Prime Minister and UN Special Envoy for Global Education and UN Education Commission Chair, Gordon Brown, summarised the criticality of financing the building of strong education systems stating that:

> Delivering an education to all – and not just some children – is the civil rights struggle of our time. Confronted by the largest refugee crisis since the close of the Second World War, and with education receiving less than 2% of humanitarian aid, it is vital we marshal the funds to provide an education for all children – especially those left out and left behind: refugee children.[20]

A welcome outcome of the meeting was that the European Union agreed to increase its support for education 'by dedicating 8 per cent of its humanitarian budget to education in emergencies in 2018, way above the global average of 3.6 per cent'.[20] Moreover, Denmark, Dubai and the IT company Hewlett Packard also committed additional funding for the 'Education Cannot Wait' fund.

Education for the twenty-first century

Closely aligned with the high-level event 'Financing the future: Education 2030', the thirty-ninth session of the UNESCO General Conference held on 10 November 2017 confirmed UNESCO's mission 'to build peace, eradicate poverty and drive sustainable development',[21] building on the 'Incheon Declaration for Education 2030'[11,12] and the '2017/8 GEM Report',[13] issued in July 2017.

Moreover, the conference underscored its 'mandate to cover all aspects of education' leading to 'the Global Education 2030 Agenda through Sustainable Development Goal 4': to 'ensure inclusive an equitable quality education and promote lifelong learning opportunities for all'.[3]

Achieving SDG 4 and seven targets plus three sub-targets will 'require increasing efforts, especially in sub-Saharan Africa and Southern Asia and for vulnerable populations, including persons with disabilities, indigenous people, refugee children and poor children in rural areas'.[22] Audrey Azoulay, former French Minister of Culture and Communication, was sworn in as the new Director-General of UNESCO on 10 November 2017.[21] In her investiture speech, Azoulay

> highlighted the challenges posed to the world today by environmental degradation, terrorism, attempts to discredit scientific findings, attacks on cultural diversity, the oppression of women and the massive displacement of

populations, saying, “I believe in the need for concerted strategies in the framework of multilateralism to face these challenges today and tomorrow”.[21]

Investing in inclusive quality education

A subsequent report in April 2018 for the G20 Education Working Group, drafted by Dr Manos Antoninis, director of the Global Education Monitoring Report and Dr Jordan Naidoo, director of UNESCO Education, addressed two key questions: ‘How can the G20 promote articulated management of international financing mechanisms?’ and ‘How can the G20 leaders ensure sufficient efficient investment for inclusive quality education?’.[23]

The report highlighted that education is not only a fundamental human right but ‘is in fact a human right which enables the realisation of other human rights’.[23] As such, it ‘reduces poverty, boosts job opportunities, and fosters economic prosperity’. Education also ‘increases people’s chances of leading a healthy life, deepens the foundations of democracy, and changes attitudes to protect the environment and empower women’.[23]

While there is growing awareness that education benefits people’s lives, there are also concerns that there is considerable under-investment ‘especially taking into account the costs of achieving the international targets in low and lower middle income countries’.[23] The report’s authors ‘estimate that annual global spending on education is approximately US$4.7 trillion’.[23] Most of these funds – US$3 trillion or around 65 per cent of the total – ‘is spent in high income countries and just US$22 billion (or 0.5% of the total) in low income countries despite the fact that the two country income groups account for a roughly equal total population of school age children.[23] Globally, the report also highlights that ‘one in four countries ... spent less than 4% of GDP on education *and* allocated less than 15% of total public expenditure to education’.[23] According to the authors, a fundamental solution to the problems lies with increases in the tax base (there are ‘too many tax incentives’) as well as tackling tax evasion, tax avoidance, corruption and ‘illegal mispricing practices’.[23]

As discussed in Chapter 2, the 2015 ‘Addis Ababa Action Agenda’ that underscores the SDGs covers all sources of finance, public, private, international development and trade – including education. With regard to external financing, ‘developed countries committed again to allocate 0.7% of gross national income to aid.’[23] It is noteworthy, considering that ‘the total [education] gap is equivalent to 8% *of annual global military expenditure*’ [italics added], that

> the estimated annual education financial gap of US$39 billion would be filled if all DAC [Development Assistance Committee] and selected donors spent (i) 0.7% of their gross national income on aid and (ii) 10% of their aid portfolio on basic and secondary education.[23]

The authors of the report[23] prepared a number of recommendations for consideration by the G20 Education Working Group. These emphasised, among other things, country commitments to SDG 4, tax policies, south–south cooperation in education, innovative financing (as in the health sector), taking a long-term perspective and calling for 'stronger complementarity and collaboration'[23] with regard to the three main education financial mechanisms: the 'Global Partnership for Education' (low- and lower-income national strategies), 'Education Cannot Wait' (emergencies and humanitarian) and the 'International Facility for Financing Education' (investment bank facility).

Preparing learners for the twenty-first century

OECD Education 2030

In the Foreword to the OECD 2030 position paper, 'The future of education and skills 2030'[24,25,26] Andreas Schleicher, OECD Director for Education and Skills, highlights that in light of the uncertainties facing the world

> students will need to develop curiosity, imagination, resilience and self-regulation; they will need to respect and appreciate the ideas, perspectives and values of others; and they will need to cope with failure and rejection, and to move forward in the face of adversity. Their motivation will be more than getting a good job and a high income; they will also need to care about the well-being of their friends and families, their communities and the planet.[24]

In response, the OECD has launched 'The Future of Education and Skills 2030' project 'to help countries find answers to two far-reaching questions:

- What knowledge, skills, attitudes and values will today's students need to thrive and shape their world?
- How can instructional systems develop these knowledge, skills, attitudes and values effectively?'.[24]

The overarching shared vision of the project is 'helping every learner develop as a whole person, fulfil his or her potential and help shape a shared future built on the well-being of individuals, communities and the planet'.[24] The project is divided into two phases. Phase 1 (2015–2018) involves the development of a conceptual learning framework relevant for 2030 and an international curriculum analysis; Phase 2 (2019 and beyond) focuses on 'building common ground on the principles and instructional designs that can effectively implement intended curricula and exploring the types of competencies and profiles of teachers who can support all students to achieve desired outcomes for their future success'.[24]

In a quest to find 'new solutions in a rapidly changing world',[24] 'The Future of Education and Skills 2030' project underlines four fundamental needs that are crucial in addressing environmental, economic and social issues. These are the need for:

- 'broader education goals' that emphasise 'individual and collective well-being';
- students 'to exercise agency, in their own education and throughout life', implying 'a sense of responsibility to participate in the world and, in so doing, to influence people, events and circumstances for the better';
- 'a broad set of knowledge, skills, attitudes and values in action';
- 'competencies to transform our society and shape our future,' in particular, 'creating new value, reconciling tensions and dilemmas and taking responsibility'.

OECD Education 2030 works with countries, thought leaders, experts, school networks, school leaders, teachers, students and social partners, and aims to help stakeholders within the education system to effectively implement curriculum reform. 'The Future of Education and Skills 2030' project allows stakeholders to exchange ideas, learn from and compare best practice and contribute to creating a new learning ecosystem. OECD Education 2030 focuses on secondary education as a starting point, while recognising the importance of the learning progression across grade levels and lifelong learning.

The working group of 'The Future of Education and Skills 2030' project is 'collecting ideas and examples of good practice for making the learning framework actionable'.[24] (For further information, see www.oecd.org/edu/school/education-2030.htm or contact education2030@oecd.org.)

The Commonwealth Education Policy Framework

Purpose and application[25]

The Commonwealth Education Policy Framework (CEPF) results from several developments: the UN-2030 SDGs and the 19th Conference of Commonwealth Education Ministers (19CCEM)[27] that drew both on the SDGs and the Incheon Declaration Framework for Action. The CEPF complements existing international frameworks and addresses some of the gaps inherent in the existing national policies of member countries. It seeks to provide the 53 Commonwealth member countries with a comprehensive approach, including to:

- identify key policy gaps and challenges;
- assist renewal and development of national education policies and legislation that enables the delivery of the SDGs in line with Commonwealth needs, priorities and values;
- ensure delivery on the critical aspects of Goal 4 of the SDGs;
- enable policy development that addresses simultaneously the challenges of equity and quality in a sustainable way and enhances the relevance of education across all SDGs.

Offering 'a series of thinking tools for considering what makes sense in specific national contexts',[27] the CEFP presents a view of how a Commonwealth perspective, embedded in Commonwealth values and supported by Commonwealth collaboration, can enrich other international and national policy processes. In particular, the CEPF, 'which will encompass a life-course perspective, takes a systems strengthening approach that will build capacity for education policy and frameworks across the Commonwealth in an effort to improve educational outcomes reflective of the SDGs.'[27]

Commonwealth Education Ministers at 19CCEM 'reaffirmed the Commonwealth's values of equity, access and development, stressing the key role of Education for Sustainable Development'.[27] They also recognised that: 'Whilst advances have been made by Commonwealth countries in education over the past 15 years, in line with the MDGs and EFA initiatives, there are still challenges that must be addressed urgently.'[27] Of these, although 'the more resourced countries were close to Universal Primary Education (UPE)' – as evidenced in the '2017/8 GEM Report' – 'only one-third of the children reached UPE in the developing countries' and 'approximately 17 million primary aged children' are out of school with 'a similar pattern for lower secondary enrolment where there are 16 million youth out of school'.[27] In the Commonwealth nations there 'are over 400 million illiterate adults' and 'approximately 3 per cent more boys than girls enrolled in primary schools'.[27]

Figure 10.2 provides an overview of the CEPF, which incorporates:

- **four cornerstones**, that is, 'the guiding principles that drive a Commonwealth view of what constitutes a good education system: quality, equitable access, relevance and sustainability';[27]
- **four enablers**: these are aspects of processes and systems that are crucial to effective policies – governance, knowledge, advocacy and capacity; and
- **six elements**: these are educational sub-systems: early childhood care and education; primary schooling; secondary schooling; technical and vocational education and training; tertiary education; and adult education and learning.

The respective components under each theme are intended to provide a high-level comprehensive overview of the main policy prescriptions required for supporting the SDGs and improving educational outcomes. Each of these is clearly elaborated in the CEPF. In respect of enablers, two may merit special attention: advocacy and capacity.

Commonwealth Ministers recognise that 'education is a complex human system that is based on a negotiated set of values' and that 'successful delivery on SDG4 will require multidirectional processes of communication between the state and other stakeholders'.[28] Their perspectives are set out in section 3.1 ('Leadership and ethics') of the CEPF, stressing the importance of shared leadership and responsibility:

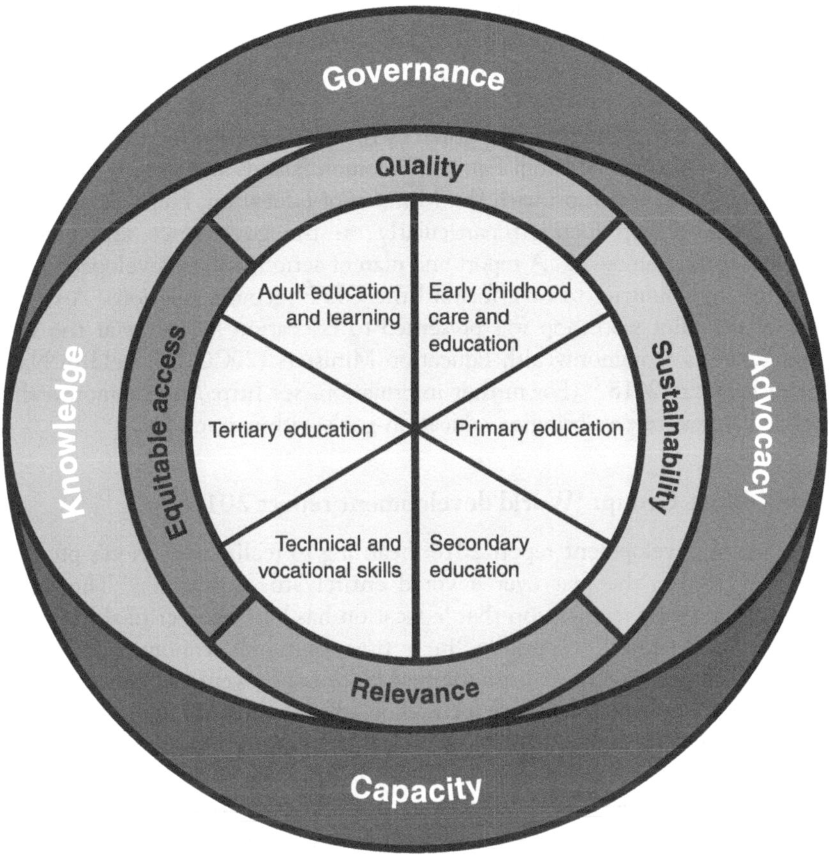

Figure 10.2 Commonwealth education policy framework.
Source: Commonwealth Secretariat.

> We have moved away from the days in which educational policymaking was seen as a top-down process in which ministers decided what needed doing; officials determined how this should be done; and other actors were merely supposed to implement. Now, the place of other stakeholders is accepted as crucial, whilst many systems have undergone decentralisation, with education often a concurrent function of different levels of government.[27]

Priorities for policy implementation include:

- 'programmes of professional development on implementation, including both planning and monitoring';

- 'the availability of technical assistance partners with extensive experience of implementation';
- 'a strong focus on the financing of implementation'; and
- 'a commitment to ongoing mechanisms for stakeholder dialogue'.[27]

In October 2017, the CEPF was piloted by the Commonwealth Secretariat with senior education officials from Fiji, the Solomon Islands and Tuvalu at a workshop held in collaboration with the Ministry of Education, Heritage and Arts, Fiji. The workshop focussed particularly on the governance and capacity enablers of the framework. A report and plan of action will be developed for the participating countries to take forward the CEPF recommendations. An overview of the pilot workshop was presented to Education Ministers at the 20th Conference of Commonwealth Education Ministers (20CCEM) held in Nadi, Fiji in February 2018.[29] (For further information, see http://thecommonwealth.org/media/press-release/bringing-education-goals-within-reach.)

World Bank Group: 'World development report 2018'

The 'World development report 2018: Learning to realize education's promise' (2018 WDR) is 'the first ever devoted entirely to education'.[30] The report represents growing recognition that 'education has long been critical to human welfare' (well-being?), especially 'in a time of rapid economic and social change'.[30] In contrast to at least the first half of the twentieth century, when 'research on learning was mainly focused on what the teacher did', it is reassuring that the pendulum has now swung to recognising that the best way to prepare children and youth for the future is 'to place their learning [at] the center'.[30] This shift is more than semantic as it recognises that 'what the student does is actually more important in determining what is learned than what the teacher does' and implies 'a need to change our beliefs about the teacher's role in learning'.[31] This re-orientation may have considerable impact on how students learn, their motivation and 'the social context of learning'.

According to Jim Yong Kim, President of the World Bank Group, 'education powerfully advances the World Bank Group's twin strategic goals: ending extreme poverty and boosting shared prosperity.'[30] Moreover, he underscores that: 'Given that today's students will be tomorrow's citizens, leaders, workers, and parents, a good education is an investment with enduring benefits.'[30] The main messages of the 'World development report 2018' are that

> in many countries and communities learning isn't happening. Schooling without learning is a terrible waste of precious resources and of human potential. Worse, it is an injustice. Without learning, students will be locked into lives of poverty and exclusion, and the children whom societies fail the most are those most in need of a good education to succeed in life. Learning conditions are almost always much worse for the disadvantaged, and so are learning outcomes. Moreover, far too many children still aren't

even attending school. This is a moral and economic crisis that must be addressed immediately.[30]

Equally concerning is that: 'This learning crisis is widening social gaps instead of narrowing them. Young students who are already disadvantaged by poverty, conflict, gender or disability reach young adulthood without even the most basic life skills.'[30]

Addressing both the economic and moral failures of education systems, the report's authors – with extensive contributions from stakeholders across the globe – explored four main themes: (1) education's promise; (2) the need to shine a light on learning; (3) how to make schools work for learners; and (4) how to make systems work for learning. Special emphasis is given to five 'spotlight' topics: the biology of learning; the impact of poverty on biological development; the multidimensionality of skills; the learning process; and implications of technology on learning. The 'World development report 2018' traces the causes of the current global learning crisis to 'poor learning outcomes'[30] and identifies five immediate conditions or causes that impede effective learning:

- children arrive unprepared to learn;
- teachers often lack the skills or motivation to teach effectively;
- inputs often fail to reach classrooms or to affect learning;
- poor management and governance often undermine schooling quality; and
- deeper systemic causes.

According to the report, recommended policy actions encompass strengthening assessment of learning, acting on evidence, placing learners at the centre and improving coordination of decision-making at all levels 'to make the system work for learners'.[30] (For further information, see www.worldbank.org/en/publication/wdr2018.)

Global educational priorities for the early decades of the twenty-first century

The UN, OECD, Commonwealth and World Bank education initiatives outlined in this chapter are indicative of a growing and shared understanding that SDG 4, 'the quality of education', underpinned by SDG 3, 'health and well-being',[2] remains the most important lever for societal transformative change, thereby laying the foundation for a global sustainable future – people and planet. In this regard, two of the most important global priorities are, unquestionably, as follows: first, ensuring that 263 million children and youth have access to education;[7] and, second, that national policy-makers, communities and educators[12,13,14] should give full attention to the quality of the student *learning* experience. More specifically, as shown in Box 10.2, three strategic reforms are recommended:

Box 10.2 Educational reforms

- Make *learner* developmental needs, present and future – both academically and socially – the main concern across all regions.
- Address systemic and pedagogical challenges across all SDG 4 targets – particularly *tackling gender inequalities* (e.g. primary school participation); prioritising provision for the *training of teachers*; *increasing higher education participation rates* based on national and local needs; and *funding* to ensure effective learning environments.
- *Strengthen instructional systems* (organisationally, aims and methods) to function effectively and efficiently *for* learners.

Along with the impact that education has on 'economic growth, employment and earnings' as well as 'helping individuals escape poverty', as discussed previously in relation to global education financing, the authors of UNESCO's '2013/2014 education for all global monitoring report'[32] stress the important role education plays in improving understanding of politics. They note, for example, that 'Across 18 sub-Saharan African countries, those of voting age with secondary education were two times as likely to express support for democracy as those with no education' and that 'education also helps overcome gender biases in political behaviour to deepen democracy'.[32] In addition, 'In India, reducing the gender literacy gap by 40% increased the probability of women standing for state assembly election by 16% and the share of votes that they received by 13%.'[32] Moreover, the relationship between education inequality and conflicts is also clear given that: 'In 55 low and middle income countries where the level of educational inequality doubled, the probability of conflict more than doubled, from 3.8% to 9.5%.'[23]

Learning in the twenty-first century

Beyond meeting immediate academic priorities there is also an increasing need to re-examine the fundamental aims and approaches of education, as evidenced by the OECD Education 2030 project.[24,25,26] The challenges are considerable for both the more economically developed as well as developing nations, especially at a time when educational systems are faced with a 'new era of Internet-powered innovation and machine learning, robots and software algorithms are replacing routine jobs traditionally done by people'.[33]

According to Kyle Pearce, an education futurist, what separates us from machines and makes human beings unique is that 'we are naturally curious, creative and social'.[33] And, 'to remain indispensable in a knowledge-based economy,' he contends, we need to focus on developing our 'sensory awareness, creativity, and social intelligence'.[33] A first step for those seeking to 'think deeply'[33] about the future of education is to consider several key questions about the purpose of education in the twenty-first century, including:

1 What does it mean to be well-educated ... ?
2 How does one live a creative, purpose-driven life that matters?
3 How do we nurture curiosity and wonder in ourselves and our children?
4 How do we inspire self-motivated learners able to solve their own problems?
5 How do we empower citizens who actively serve their communities?[33]

Similar queries have been raised at the Harvard Graduate School of Education where interdisciplinary learning is a core feature of Harvard's Project Zero (PZ) founded at the School in 1967.[34] The project 'began investigating aspects of cognitive processing in the arts' and expanded its insights 'to include broader aspects of human potential such as intelligence, learning, critical thinking, and creativity'.[34] Driving its investigations are 'fundamental questions of human potential as they relate to contemporary issues facing an array of educational settings, such as schools, families, museums, and businesses'.[34]

Veronica Boix Mansilla, who chairs the Future of Learning Institute and directs the Interdisciplinary and Global Studies Project, has described the characteristics of learners in the twenty-first century:

> Underlying the significant social and technical changes of our increasingly global society is a profound transformation in the very nature of learning. Twenty-first-century learners view themselves as dynamic agents in multimedia and global environments. These learners create complex social networks and operate comfortably in them. They learn experientially in real and virtual worlds alike. They express their views and make their lives public with ease. Many of these learners show greater interest in the global environment and human rights than their immediate predecessors.[35]

With the need to overcome 'last century's knowledge fragmentation', Boix Mansilla's main focus is on the importance of integrative learning, the development of global competence – 'a global self' – including nurturing 'the capacity and disposition to understand and act on issues of global significance'.[35] She echoes the transformation in education advocated by faculty members at the University of Michigan, summarised below,[36] and calls for learning that is 'deep, relevant and long-lasting' and that prepares learners for 'the new global economy, a new digital world' and 'a world of complex demographics'.[35]

> We believe that the major problems of our time, from the environment to poverty, from human rights to terrorism, from religious movements to healthcare, cannot be studied effectively within any single discipline; all involve integrative thinking ... In order to prepare for a life of productive endeavour in the 21st century, [students] must learn problem solving across disciplines and launch inquiries in uncharted territories of knowledge and practice.

> They must examine the assumptions that inhere in a disciplinary perspective and integrate material outside the pattern of thought … And they must learn how to find their way through disconnected bodies of information and perspectives and create their own path to an education that coheres.[36]

A major strength of this re-orientation is that 'we can legitimately ask in class' where 'our world of today and tomorrow becomes a source of problems for study and helping students to make productive sense of such problems'.[36]

Perhaps, above all, as UNESCO reminds us:

> It is not enough for education to produce individuals who can read, write and count. Education must be transformative and bring shared values to life. It must cultivate an active care for the world and for those with whom we share it … It requires transforming the way people think and act.[37]

Finally, it is certain that implementing SDG Target 4.7 across all education sectors and levels could make a fundamental contribution towards achieving a 'more just, sustainable and peaceful world'.[38]

> By 2030, ensure that all learners acquire the knowledge and skills needed to promote sustainable development, including, among others, through education for sustainable development and sustainable lifestyles, human rights, gender equality, promotion of a culture of peace and non-violence, global citizenship and appreciation of cultural diversity and of culture's contribution to sustainable development.[39]

References

1 United Nations. *Goal 16: Promote just, peaceful and inclusive societies*. www.un.org/sustainabledevelopment/peace-justice/ (accessed 6 March 2018).
2 United Nations. *Goal 3: Ensure healthy lives and promote well-being for all at all ages*. www.un.org/sustainabledevelopment/health/ (accessed 6 March 2018).
3 United Nations. *Goal 4: Ensure inclusive and quality education for all and promote lifelong learning*. www.un.org/sustainabledevelopment/education/ (accessed 6 March 2018).
4 United Nations. *Global development goals: Leaving no one behind*. www.una.org.uk/global-development-goals-leaving-no-one-behind (accessed 29 February 2013).
5 Tombs R. Liberals are undermining western civilisation. *The Times*, 23 April 2018: 22.
6 United Nations. *The millennium development goals report 2015*. www.un.org/millenniumgoals/2015_MDG_Report/pdf/MDG%202015%20rev%20(July%201).pdf (accessed 6 March 2018).
7 Global Education Monitoring Report. *Reducing global poverty through universal primary and secondary education* [GEMR/UNESCO Institute for Statistics Policy Paper 32/Fact Sheet 44, June 2017]. http://unesdoc.unesco.org/images/0025/002503/250392E.pdf (accessed 12 May 2018).
8 UNESCO Institute of Statistics. *More than one-half of children and adolescents are not learning worldwide* [Fact Sheet 42, September 2017]. http://uis.unesco.org/sites/default/

files/documents/fs46-more-than-half-children-not-learning-en-2017.pdf (accessed 12 May 2018).

9 UNESCO. *263 million children and youth are out of school from primary to upper secondary*. www.unesco.org/new/en/education/themes/leading-the-international-agenda/education-for-all/single-view/news/263_million_children_and_youth_are_out_of_school_from_primar/ (accessed 25 September 2017).

10 UNICEF. *Syrian refugees and other affected populations in Turkey, Lebanon, Jordan, Iraq and Egypt*. www.unicef.org/appeals/syrianrefugees.html (accessed 10 January 2018).

11 UNESCO, UNDP, UNFPA, UNHCR, UNICEF, UN Women, World Bank Group. *Education 2030. Incheon declaration and framework for action. Towards inclusive and equitable quality education and lifelong learning for all*. www.unesco.org.pk/education/documents/2015/SDG-4/Framework_for_Action_Education_2030.pdf (accessed 10 June 2015).

12 UNESCO. *SDG 4 – Education 2030 framework for action*. http://unesdoc.unesco.org/images/0024/002456/245656E.pdf (accessed 20 January 2016).

13 UNESCO. *Global monitoring report 2017/8. Accountability in education: Meeting our commitments*. http://unesdoc.unesco.org/images/0025/002593/259338e.pdf (accessed 22 July 2017).

14 Lueddeke GR. Development of a project proposal supporting a sustainable future for people and planet. *Africa Health News*, 30 May 2017. https://africahealthnews.com/development-project-proposal-supporting-sustainable-future-people-planet/ (accessed 20 October 2017).

15 Møller M, Roberts BJ, Tiliouine H, Loschky J. Waiting for happiness in Africa [*World happiness report 2017*, Chapter 4]. https://s3.amazonaws.com/happiness-report/2017/HR17-Ch4_w-oAppendix.pdf (accessed 29 June 2018).

16 Naidoo, J. *Agenda 2030 – SDG 4 Education 2030 – one year on: Challenges and opportunities*. www.norrag.org/agenda-2030-sdg4-education-2030-one-year-on-challenges-and-opportunities/ (accessed 29 June 2018).

17 Mead L. SDG report 2017 provides overview of efforts to achieve SDGs. *International Institute for Sustainable Development*, 27 July 2017. http://sdg.iisd.org/news/sdg-report-2017-provides-overview-of-efforts-to-achieve-sdgs/ (accessed 28 March 2017).

18 UNESCO. *Target 4.7 – Sustainable development and global citizenship*. http://gem-report-2017.unesco.org/en/chapter/target-4-7-sustainable-development-and-global-citizenship-2/ (accessed 25 July 2017).

19 UNICEF. World leaders commit to tackling global education crisis that is holding back millions of children and threatening progress and stability [news release, 20 September 2017]. https://en.unesco.org/news/world-leaders-commit-tackling-global-education-crisis-holding-back-millions-children-and (accessed 25 September 2017).

20 Global Partnership for Education. *Financing the future: Education 2030*. www.globalpartnership.org/event/financing-future-education-2030 (accessed 26 September 2017).

21 UNESCO. UNESCO more necessary than ever, declares Audrey Azoulay [news release, 13 November 2017]. https://en.unesco.org/news/unesco-more-necessary-ever-declares-audrey-azoulay (accessed 21 November 2017).

22 United Nations. Progress of goal 4 in 2017. https://sustainabledevelopment.un.org/sdg4 (accessed 20 November 2017).

23 Naidoo J, Antoninis M. *The status and architecture of global education financing: A background paper for the 2018 G20*. www.bundesfinanzministerium.de/Content/EN/Standardartikel/Topics/Featured/G20/G20-Finance-Track-Documents.html (accessed 10 May 2018).

24 OECD. *The future of education and skills: Education 2030* [position paper, 2018]. www.oecd.org/education/2030/oecd-education-2030-position-paper.pdf (accessed 20 July 2018).

25 Schleicher A. *The case for 21st-century learning*. www.oecd.org/general/thecasefor21st-centurylearning.htm (accessed 28 October 2017).

26 OECD. *Education 2030*. www.oecd.org/edu/school/education-2030.htm (accessed 25 September 2017).
27 The Commonwealth. *Commonwealth education policy framework*. www.thecommonwealth-educationhub.net/policyframework/ (accessed 27 February 2017).
28 The Commonwealth Secretariat. *Quality education for equitable development: Performance, paths and productivity* [19th Conference of Commonwealth Education Ministers, Nassau, The Bahamas, 22–26 June 2015]. www.thecommonwealth-educationhub.net/wp-content/uploads/2016/09/19CCEM-ReportFinal.pdf (accessed 27 July 2015).
29 The Association of Commonwealth Universities. Education can deliver: Ministers recognise importance of higher education at 20CCEM [news release, 26 February 2018]. www.acu.ac.uk/news/view?id=203&x[0]=news/list (accessed 25 February 2018).
30 World Bank. *Learning to realize education's promise* [World development report 2018]. https://openknowledge.worldbank.org/bitstream/handle/10986/28340/211096ov.pdf (accessed 20 February 2018).
31 Biggs JB. *Teaching for Quality Learning at University*. 2nd ed. Buckingham, UK: Open University Press/Society for Research into Higher Education; 2003.
32 UNESCO. *Teaching and learning: Achieving quality for all* [EFA global monitoring report 2013/4]. http://unesdoc.unesco.org/images/0022/002266/226662e.pdf (accessed 27 February 2018).
33 Pearce K. *21st century education: Re-imagining learning for a new era*. www.diygenius.com/21st-century-education/ (accessed 27 February 2018).
34 Project Zero. *Who we are*. www.pz.harvard.edu/who-we-are (accessed 25 September 2017).
35 Boix Mansilla V. *Integrative learning: Setting the stage for a pedagogy of the contemporary*. www.pz.harvard.edu/resources/integrative-learning-setting-the-stage-for-a-pedagogy-of-the-contemporary (accessed 24 January 2018).
36 Gnagey LT. *Teaching initiative funding announced*. www.ur.umich.edu/0708/Sept04_07/09.shtml (accessed 25 February 2018).
37 UNESCO. *Global education first initiative. Priority #3: Foster global citizenship*. www.unesco.org/new/en/gefi/priorities/global-citizenship (accessed 29 November 2017).
38 United Nations. *Transforming our world: the 2030 agenda for sustainable development*. https://sustainabledevelopment.un.org/post2015/transformingourworld (accessed 20 September 2017).
39 United Nations. *Measurement strategy for SDG target 4.7: Proposal by GAML task force 4.7*. http://uis.unesco.org/sites/default/files/documents/gaml4-measurement-strategy-sdg-target4.7.pdf (accessed 27 January 2018).

11 Collaborative *One Health and Well-Being* initiatives

Introductory comments and chapter purpose

Chapters 8 and 9 provided background information on the evolution of *One Health*, tracing its origins from Hippocrates and medical practice in ancient Greece, when doctors realised that diseases have natural rather than spiritual causes,[1] to the multidisciplinary and multiprofessional approaches to complex issues we are adopting today – particularly relating to the physical and social interconnections between human, animal, plant and ecological health and well-being. The *One Health* concept – which began as an approach and became a global movement with an emerging philosophy and set of values and principles[2] – has been defined as 'a worldwide strategy for expanding interdisciplinary collaborations and communications in all aspects of health care for humans, animals and the environment'.[3]

The *One Health* movement's main aim is to 'improve *health and well-being* through the prevention of risks and the mitigation of effects of crises that originate at the interface between humans, animals and their various environments'.[4] Countries are increasingly placing a very high premium on the integration (vs fragmentation) of approaches to sustainability, exemplified by the UN-2030 Global Goals or SDGs and the hundreds of global and national organisations that have endorsed *One Health*. While there are differences in emphases, there is a growing consensus around the necessity to adopt a new mindset and make a paradigm shift, replacing the current view of limitless resources and our huge appetite for exploitation, competition and conflict with one that respects the sanctity of life and strives towards shared wealth, well-being, prosperity and social stability.[5]

With the latter worldview in mind, the main purpose of Chapter 11 is to disseminate developments by leading global and national organisations that share a commitment to sustaining planet Earth – people and planet. The contributions (including in some cases associated references), have been organised alphabetically and were kindly submitted by individuals working on groundbreaking *One Health and Well-Being* initiatives. Credit for their contributions is given in the Acknowledgements.

The following organisations have contributed to this chapter:

1 Centers for Disease Control and Prevention (CDC)
2 Chatham House
3 The EcoHealth Alliance
4 The Institute of Global Health, Faculty of Medicine, University of Geneva
5 The InterAction Council
6 The International Veterinary Students' Association (IVSA)
7 The Network for Evaluation of One Health (NEOH)
8 The One Health Commission (OHC)
9 The One Health Initiative (OHI)
10 The One Health Platform (OHP)
11 The One Health Tripartite collaboration (FAO, OIE and WHO)
12 The Planetary Health Alliance (PHA)
13 The Southeast Asia One Health University Network (SEAOHUN)
14 The United Nations Development Programme (UNDP)
15 The World Bank Group (WBG)
16 The World Veterinary Association (WVA) and World Medical Association (WMA) One Health collaboration
17 The World Wildlife Fund (WWF)

References

1 Lueddeke GR. *Global Population Health and Well-Being in the 21st Century: Toward New paradigms, Policy and Practice*. New York: Springer Publications; 2016.
2 One Health Commission. *World health through collaboration*. www.onehealthcommission.org/ (accessed 20 February 2018).
3 One Health Initiative. *One Health Initiative will unite human and veterinary medicine*. www.onehealthinitiative.com/ (accessed 20 February 2018).
4 Gibbs, E, Paul J. *The evolution of One Health: A decade of progress and challenges for the future*. https://veterinaryrecord.bmj.com/content/174/4/85 (accessed 20 February 2015).
5 Rogers DS, Duraiappah AK, Antons DC, Munoz P, Bai X, Fragkias M, *et al*. A vision for human well-being: Transition to social sustainability. *Current Opinion in Environmental Sustainability* 2012; 4: 1–13.

1 Centers for Disease Control and Prevention (CDC)

Operationalising a One Health Approach at the US Centers for Disease Control and Prevention

Karen Alroy, One Health Lead and Casey Barton Behravesh, Director, One Health Office, National Center for Emerging and Zoonotic Infectious Diseases

Overview

The US Centers for Disease Control and Prevention (CDC) recognise that the health and well-being of people is connected to the health of animals and our shared environment. This core principle was put into action near the beginning of CDC's history. Dr James H. Steele, an early *One Health* advocate, created the Veterinary Public Health Division at CDC in 1947 in recognition of the important role that animals play in public health. In 2009, building on this foundation, the CDC established a *One Health* office,[1] a team within the National Center for Emerging and Zoonotic Infectious Diseases (NCEZID), responsible for leading and coordinating *One Health* activities within CDC in concert with other domestic and international agencies and organisations. As the lead public health agency for the US, CDC's mission is to protect the health of Americans where they live, work and play.

While CDC strives towards this mission in a variety of ways, utilising a *One Health* approach and collaborating with relevant human, animal and environmental health experts and stakeholders has been found to be critical in achieving this goal. CDC has a leadership role in a number of important *One Health* issues such as emerging zoonotic and infectious diseases, antimicrobial resistance and food safety. CDC laboratory and public health scientists investigate zoonotic and emerging infectious disease threats. These scientists seek to understand transmission dynamics between humans, animals and their shared environment behaviours and risk factors that may influence whether a person becomes sick from an animal or environmental exposure, the speed at which these diseases can spread and how to best prevent these illnesses.

CDC supports the detection, investigation and response to events and public health emergencies within the US and around the world and collaborates with *One Health* partners. For example, during several major public health events, such as the West Nile virus introduction into the US in 1999, the anthrax bioterrorist events in 2001, the 2014–2016 Ebola outbreak in West Africa and the 2015 Zika virus outbreak in the Americas, CDC has been an important *One Health* partner addressing these complex public health threats that impact the health of Americans and others around the world. CDC subject matter experts develop public health policies and interventions regarding public health issues at the human–animal–environment interface.

CDC also works closely with partners to coordinate and communicate these important messages to the public.

Initiatives

CDC is the lead federal agency for zoonotic diseases in people in the US. CDC's *One Health* Office has a leadership role in strengthening a *One Health* approach in the US. For example, in 2017 CDC, in collaboration with the US Department of Agriculture and the Department of the Interior, hosted a *One Health* Zoonotic Disease Prioritisation Workshop for the US. This *One Health* workshop had two goals: (1) to use a *One Health* approach to prioritise the zoonotic diseases of greatest national concern that should be jointly addressed by human, animal and environmental health agencies; and (2) to develop plans for implementing and strengthening multisectoral *One Health* approaches to address these diseases in the US. Participants representing federal and state human, animal and environmental health agencies identified eight priority zoonotic diseases requiring a multisectoral, *One Health* approach for control and prevention: zoonotic influenza viruses, salmonellosis, West Nile virus, plague, emerging coronaviruses, rabies virus, brucellosis, and Lyme disease. The workshop was a critical step towards a unique US approach to *One Health*, ensuring that all stakeholders have a shared vision and roadmap for implementing *One Health* for disease surveillance, response, preparedness and prevention and control activities in their current and future areas of focus. CDC also has a leadership role in preventing zoonotic diseases shared between people and pets. The CDC *One Health* Office leads the 'Healthy Pets, Healthy People'[2] programme to provide education on how to enjoy animals without getting sick and reducing the risk of zoonotic disease transmission between people and pets and other animals such as livestock and wildlife.

The *One Health* work that CDC carries out extends beyond US national boundaries, and often involves collaboration and coordination with international partners to promote global health security including preventing shared health threats at the human–animal–environment interface. The CDC *One Health* Office has taken a strategic role in helping countries identify their priority zoonotic diseases of greatest national concern for *One Health* collaboration. Through *One Health* Zoonotic Disease Prioritisation workshops, CDC trained facilitators have led government stakeholders – from the human, animal and environment sectors – to use a semi-quantitative tool developed at CDC to identify their country's top zoonotic diseases of greatest national concern.[3,4,5] This process not only identifies priority zoonotic diseases using an established, transparent methodology, but it allows countries to use a *One Health* approach to develop action plans to address the newly prioritised diseases in order to build and strengthen important multisectoral relationships for advanced *One Health* coordination. As of April 2018, 20 countries in Africa, Asia, Europe and the Americas have conducted a One Health Zoonotic Disease Prioritisation Workshop. These countries are seeing the benefits of using a collaborative, multisectoral, *One Health* approach to prevent and control zoonotic diseases.

CDC also assists countries in developing National *One Health* strategic plans that incorporate the priority zoonotic diseases as well as other *One Health* issues of national concern, and outlines action plans and roles and responsibilities for all sectors engaged in *One Health* in a country. To help train the next generation of *One Health* professionals, CDC plays an important role in workforce development in the US through its Epidemic Intelligence Service (EIS) programme and around the world through the Field Epidemiology Training Programme (FETP) hosted in more than 60 countries.

These *One Health* global activities and collaborations are underpinned by a strong relationship between the CDC and other global *One Health* leaders. Coordination and collaboration with the tripartite World Health Organization (WHO), World Organisation for Animal Health (OIE) and the Food and Agriculture Organization of the UN (FAO), as well as other global One Health partners, has improved the impact and effectiveness of CDC's *One Health* work around the world. This type of collaboration is highlighted by CDC's designation as an OIE Collaborating Center for Emerging and Re-Emerging Zoonotic Diseases. With increasing global connectedness, and rapidly changing landscapes and livelihoods, the importance of utilising a multisectoral, One Health approach is critical for best protecting the health of people and animals and our shared environment.

References

1 CDC. *One Health.* www.cdc.gov/onehealth (accessed 20 July 2018).
2 CDC. *Healthy Pets, Healthy People.* www.cdc.gov/healthypets (accessed 20 July 2018).
3 Rist CL, Arriola CS, Rubin C. Prioritizing zoonoses: a proposed one health tool for collaborative decision-making. *PLoS One* 2014; 9(10): e109986. http://journals.plos.org/plosone/article?id=10.1371/journal.pone.0109986 (accessed 23 July 2018).
4 Salyer SJ, Silver R, Simone K, Barton Behravesh C. Prioritizing zoonoses for global health capacity building – Themes from one health zoonotic disease workshops in 7 countries, 2014–2016. *Emerging Infectious Diseases* 2017; 23(13). www.nc.cdc.gov/eid/article/23/13/17-0418_article (accessed 23 July 2018).
5 CDC. *Zoonotic Disease Prioritization.* www.cdc.gov/onehealth/global-activities/prioritization.html (accessed 20 July 2018).

For more information about the CDC, please visit www.usa.gov/federal-agencies/centers-for-disease-control-and-prevention

2 Chatham House

Operationalising One Health in the UK: the Human Animal Infections and Risk Surveillance (HAIRS) Group

Julia Spencer and Afifah Rahman-Shepherd, Centre on Global Health Security

Overview

Chatham House, the Royal Institute of International Affairs, is an independent policy institute based in London. Its mission 'is to help build a sustainably secure, prosperous and just world, through informed debate, independent analysis, new policy ideas, and outreach to audiences'.[1] In 2016, Chatham House was named 'Think Tank of the Year' at Prospect magazine's annual think tank awards, where it also won in the UK categories for International Affairs and Energy and Environment. The judges commented that the institute's work is 'reliably excellent' and a 'gold standard of knowledge and professionalism'.

Initiatives

In response to the public health threat from new and emerging infectious diseases, particularly at the human–animal interface, the Human Animal Infections and Risk Surveillance (HAIRS) group was established by Public Health England[1] in 2004 as a cross-government *One Health* Initiative (OHI).[2,3] The group – which involves stakeholders from a range of government departments and agencies, spanning the human health, animal and plant health, and environment and agricultural sectors – meets monthly to review new and emerging human and animal infections, and to assess the risk they pose to public health in the UK.[1,2] The group carries out horizon scanning, risk assessment and risk management activities.

The HAIRS group is recognised as a national focal point for identifying potential zoonotic incidents[2] and has demonstrated how a multi-agency, cross-sectoral collaboration can effectively conduct risk assessments for emerging health threats.[4] For example, in 2016 the group responded to the first human case of locally-acquired Crimean–Congo haemorrhagic fever in Western Europe by mobilising existing stakeholder groups and resources to rapidly determine the risk of infection in the UK.[1] Importantly, unlike many OHI's,[5,6] all stakeholders have an equal role in the group and the working relationships between them are developed ahead of a crisis and maintained with monthly meetings and frequent email exchanges.[2] The group exemplifies a truly integrated approach towards improving alerts, risk assessments and communications across the UK government and internationally.[2] The group has also been successful in achieving high level buy-in and support, even in the context of public sector budget cuts in the

UK.[2] The impact and sustainability of the HAIRS group demonstrates the added value of adopting a *One Health* approach to national surveillance and response activities.

References

1 Public Health England. *The Human Animal Infections and Risk Surveillance (HAIRS) group: 2016 report*. London; 2017. www.gov.uk/government/uploads/system/uploads/attachment_data/file/660444/HAIRS_Report_2016.pdf (accessed 16 July 2018).
2 Morgan D. One Health in action: the work of the HAIRS group. *Veterinary Record* 2014; 175(3): 61–63.
3 Heymann DL, Dar OA. Prevention is better than cure for emerging infectious diseases. *British Medical Journal* 2014; 348: g1499.
4 Lightfoot N, Rweyemamu M, Heymann DL. Preparing for the next pandemic. *Brititsh Medical Journal* 2013; 346: f364.
5 Mazet JAK, Uhart MM, Keyyu JD. Stakeholders in one health. *Revue Scientifique et Technique* 2014; 33(2): 443–452.
6 Binot A, Duboz R, Promburom P, Phimpraphai W, Cappelle J, Lajaunie C, *et al.* A framework to promote collective action within the One Health community of practice: Using participatory modelling to enable interdisciplinary, cross-sectoral and multi-level integration. *One Health* 2015; 1: 44–48.

For more information about Chatham House, please visit www.chathamhouse.org/about

3 The EcoHealth Alliance

A One Health approach to protecting wildlife, the environment and public health

Catherine Machalaba, Policy Advisor, Science Officer, Future Earth One Health Project; Chair, Veterinary Public Health Special Primary Interest Group, American Public Health Association

Overview

Building on over 45 years of groundbreaking science, EcoHealth Alliance is dedicated to a *One Health* approach to protecting the health of people, animals and the environment from emerging infectious diseases. Approximately 60 per cent of emerging infectious diseases such as Ebola, HIV, Zika, severe acute respiratory syndrome (SARS), Middle East respiratory syndrome (MERS) and West Nile virus have all originated in animals before spilling over to human populations. Using environmental and health data covering the past 60 years, EcoHealth Alliance scientists created the first ever global disease hotspots map that identified at risk regions to determine where research and field work are needed to help predict and prevent the next pandemic crisis. That work is the foundation of EcoHealth Alliance's rigorous, science-based approach in nearly 30 countries worldwide. EcoHealth Alliance's strength is founded on innovations in research, training, global partnerships, capacity building and policy initiatives.

Initiatives

The organisation uses a multidisciplinary method to solve health challenges caused by global changes and human–animal interactions. EcoHealth Alliance works with local governments, in-country scientists and policymakers around the world to make critical on-the-ground changes for the prediction and prevention of infectious disease. Its dedicated scientists conduct field research and develop tools to save ecosystems and predict and prevent pandemics. In addition to our extensive partners in countries at all levels of government, EcoHealth Alliance has longstanding collaborations with numerous international technical and governing bodies, including the World Bank, the UN Convention on Biological Diversity, WHO, OIE and the FAO. EcoHealth Alliance also hosts the International Union for Conservation of Nature (IUCN) Species Survival Commission Wildlife Health Specialist Group and a global research project under the sustainability research platform Future Earth.

EcoHealth Alliance partnered the development of the World Bank's 'Operational framework for strengthening human, animal and environmental public health systems at their interface' (or 'One Health operational framework'). This new framework aims to provide a basis for understanding and implementing a *One Health* approach in global and country investments. Building on the World Bank's previous 'People, pathogens, and our planet' reports, it examines the strategic context, rationale and the case for investing in *One Health*; provides an inventory of tools and other resources from the human health, agriculture, environment and disaster risk reduction sectors to facilitate multisectoral coordination; and includes technical guidance for appropriate *One Health* entry points and operations along the prevent–detect–respond–recover spectrum of preparedness.

For more information about EcoHealth Alliance, please visit www.ecohealthalliance.org

4 The Institute of Global Health, Faculty of Medicine, University of Geneva

The One Health perspective of UNIGE's Institute of Global Health: international, transdisciplinary, humanitarian and digital

Rafael Ruiz de Castañeda, Isabelle Bolon, Sara Babo-Martins, Danny Sheath, Professor Antoine Flahault

Overview

With a unique concentration of international organisations, UN permanent missions, NGOs, scientific and medical societies, public–private partnerships and academic institutions, International Geneva[1] is often seen as the world's capital for humanitarian action and global health. UNHCR, International Telecommunication Union (ITU), United Nations Environment Programme (UNEP), the Global Alliance for Vaccines and Immunisation (GAVI), Médecins Sans Frontières (MSF) and, of course, WHO are some of the major global health players based in Geneva with close links to its university and hospitals –Université de Genève (UNIGE) and Geneva University Hospitals (HUG).

UNIGE's Institute of Global Health (IGH)[2] is at the core of this international and interdisciplinary ecosystem, and its educational and research activities embrace this diversity, benefitting from active collaborations. The IGH is located at Campus Biotech,[3] which brings together experts in life and data sciences from the UNIGE and École Polytechnique Fédérale de Lausanne (EPFL) to promote digital innovation in global health and *One Health*. At the same time, the IGH is situated next to the Graduate Institute (Geneva), with outstanding expertise in social and human sciences (anthropology, global health governance, policy sciences). IGH's Precision One Health Unit (POHU) aligns fully with these dimensions being international, transdisciplinary, humanitarian and digital, thus promoting an innovative perspective to education and research on a diversity of global health issues at the human–animal–ecosystem interface.

Initiatives

First MOOC on global health at the human–animal–ecosystem interface

The IGH is an active developer of Massive Online Open Courses (MOOCs) including, for instance, MOOCs on Ebola[4] and Zika[5] during both epidemic crises in 2014 and 2016, respectively. In 2017, the IGH launched a MOOC on 'Global health at the human–animal–ecosystem interface',[6] led by the POHU and developed in collaboration with Institut Pasteur, University of Montreal and Centre Virchow-Villermé (Sorbonne, Paris). It involved 37 experts from 20 institutions, including the WHO, who presented and discussed major research findings and public health interventions for tackling zoonotic emerging infectious diseases, antimicrobial resistance, neglected tropical diseases etc. The *One Health* and 'planetary health' concepts were introduced and discussed by experts in different case studies. The content of this MOOC evolves actively to follow the global health agenda and in May 2018 (International Day for Biodiversity), a new module on 'Biodiversity and health' was launched addressing planetary health issues such as food security, climate change and health more deeply. This new module was developed in collaboration with experts from the WHO and the Convention on Biological Diversity, building on 'Connecting global priorities: Biodiversity and human health – A state of knowledge review'.[7] This MOOC is open and freely accessible to everybody worldwide.

From MOOCs to MOORs and research capacity building in One Health

MOOC platforms and many of the universities producing them tend to rely on passive models of dissemination and use of MOOCs; thus, their impact remains debated. The IGH and its POHU have explored new models for more active use of MOOCs, trying to connect their interdisciplinary and international community of experts and learners through blended learning activities and collaborative research.[8] For instance, MOOCs are integrated within UNIGE's PhD programme in Global Health and accredited by UNIGE–SSPH+ (Swiss School of Public Health[9]). In 2017, the IGH, the Swiss Tropical and Public Health Institute (TPH) and the University of Montreal organised a workshop on the 'First global flipped classroom on *One Health*: from MOOCs to real world challenges'[10] in Geneva. With the support of SSPH+, this workshop brought together 12 national and international learners from the MOOCs on 'Global health at the human–animal–ecosystem interface' (Coursera) and 'One Health: connecting humans, animals and the environment'[11] (Future Learn) to collaborate in small teams using *One Health* and innovation to tackle real world challenges: emerging infectious diseases and bushmeat trading, snakebite prevention in agricultural workers etc. Learners from various academic levels (e.g. Masters, PhD) were supported and guided by experts from International Geneva and other Swiss and international institutions, presenting potential opportunities for the publication of their work, development of ideas, funding etc. A similar model focusing on locally identified problems at the human–animal interface has recently been tested in collaboration with InZone[12] in Kakuma refugee camp, Kenya. The ultimate goal is to build *One Health* capacity among refugees and to integrate them into a larger global community of *One Health* learners and professionals via MOOCs and blended learning and research experiences.

The snakebite humanitarian crisis: opportunities for Precision One Health in the digital age

Snakebite is a silent humanitarian crisis, causing over 100,000 deaths and 400,000 disabilities every year in poor and rural communities of the world.[13] In June 2017, snakebite was included in the WHO's list of Neglected Tropical Diseases, creating unprecedented political and scientific momentum, and an urgency for innovative ideas and solutions to tackle this major global health problem. The university hospitals of Geneva together with MSF have worked on the clinical management of snakebite for over a decade and the IGH and its POHU have now brought the Precision One Health perspective to it. In the context of the Swiss National Science Foundation snakebite project,[14] the impact of snakebite will be quantified through a national household epidemiological survey in Nepal and Cameroon, looking at both the direct impact on human health and the indirect impact on livestock animals and subsequent livelihoods.

To our knowledge this is the first *One Health* approach to snakebite, including the economic dimension.[15] However, the clinical management of snakebite relies primarily on a zoological question: What species is the biting snake? The correct identification of the snake is crucial since many antivenoms are species-specific, but the clinicians usually lack the herpetological knowledge to identify the snake. With the support of Fondation Privée des HUG,[15] POHU and other collaborators are currently working on 'Snapp: first medical decision-support tool for snake identification based on artificial intelligence and remote collaborative expertise'.[16] This project will build on a recent study by the same team, showing the potential of open data sources and citizens for the collection and identification of venomous snakes across the globe.[17]

Another important area of IGH research is emerging infectious diseases and the so-called Precision Epidemic Forecasting approach. The IGH and the Geneva Centre for Emerging Viral Diseases[18] claim that through interdisciplinary development, combining digital innovations with life sciences and social and human sciences, we can transform the way we predict, detect and react to emerging infectious diseases, thus saving human and animal lives as well as economic costs.[19]

References

1 International Geneva. *Home*. www.geneve-int.ch/ (accessed 20 July 2018).
2 University of Geneva, Institute of Global Health. *Home*. www.unige.ch/medecine/isg/en/. (accessed 20 July 2018).
3 Campus Biotech. *Home*. www.campusbiotech.ch/ (accessed 20 July 2018).
4 Ebola MOOC. www.coursera.org/learn/ebola-vaincre-ensemble (accessed 20 July 2018).
5 Zika MOOC. www.coursera.org/learn/zika (accessed 20 July 2018).
6 Global health at the human–animal–ecosystem interface MOOC.www.coursera.org/learn/global-health-human-animal-ecosystem (accessed 20 July 2018).

7 UNEP, CBD, WHO. *Connecting global priorities: Biodiversity and human health – A state of knowledge review*. www.cbd.int/health/SOK-biodiversity-en.pdf (accessed 20 July 2018).

8 Flückiger Y, Achard P. From MOOCs to MOORs: a movement towards Humboldt 2.0. *Economica* 2016: 360. https://archive-ouverte.unige.ch/unige:79947 (accessed 20 July 2018).

9 Swiss School of Public Health. *Home*. http://ssphplus.ch/ (accessed 20 July 2018).

10 Ruiz de Castañeda R, Garrison A, Haeberli P, Crump L, Zinsstag J, Ravel A, *et al. First 'Global Flipped Classroom in One Health': From MOOCs to research on real world challenges*. www.ncbi.nlm.nih.gov/pmc/articles/PMC6000812/ (accessed 28 May 2018).

11 *One Health: Connecting humans, animals and the environment* [online course]. www.futurelearn.com/courses/one-health (accessed 20 July 2018).

12 InZone. *Who we are*. www.unige.ch/inzone/who-we-are/ (accessed 20 July 2018).

13 Gutiérrez JM, Calvete JJ, Habib AG, Harrison RA, Williams DJ, Warrell DA. *Snakebite envenoming*. www.ncbi.nlm.nih.gov/pubmed/28905944 (accessed 4 December 2017).

14 University of Geneva, Institute of Global Health. *One Health*. www.unige.ch/medecine/isg/en/research/environmental-health-and-health-promotion/one-health (accessed 20 July 2018).

15 Bolon I, *et al.* in preparation (Swiss National Science Foundation snakebite project – www.fondationhug.org/).

16 University of Geneva, Institute of Global Health. *Snapp: first medical decision-support tool for snake identification based on artificial intelligence and remote collaborative expertise* [project]. www.unige.ch/medecine/isg/en/research/environmental-health-and-health-promotion/one-health/snapp-first-medical-decisionsupport-tool-for-snake-identification-based-on-artificial-intelligence-and-remote-collaborative-expe/. (accessed 20 July 2018).

17 Genevieve LD, Ray N, Chappuis G, Alcoba G, Mondarini MR, Bolon I *et al.* Participatory approaches and open data on venomous snakes: A neglected opportunity in the global snakebite crisis? *PLoS Neglected Tropical Diseases* 2018; 12(3): e0006162. http://journals.plos.org/plosntds/article?id=10.1371/journal.pntd.0006162 (accessed 27 May 2018).

18 Geneva Centre for Emerging Viral Diseases· www.hug-ge.ch/en/centre-emerging-viral-diseases (accessed 20 July 2018).

19 Flahault A, Geissbuhler A, Guessous I, Guérin P, Bolon I, Salathé M, *et al.* Precision global health in the digital age. *Swiss Medical Weekly* 2017; 147: w14423. https://smw.ch/article/doi/smw.2017.14423 (accessed 16 July 2018).

For more information about IGH, please visit http://unige.g2hp.net/structure/institute-of-global-health/

5 The InterAction Council

The Dublin Charter for One Health

John Wyn Owen, Senior Advisor, Global Health and Joanna Nurse, Strategic Advisor

Overview

The InterAction Council was the first group of former statesmen to be established, initiated by political leaders in 1983.[1] It now consists of over 40 former heads of government and is presently co-chaired by President Obasanjo of Nigeria and Bertie Ahern, former Prime Minister of Ireland. The Council makes recommendations and proposals for action that are disseminated to over 1,000 people, including 300 heads of state, political leaders and UN agencies. Its main priority areas are:

- peace and security
- world economic revitalisation
- universal ethical standards.

Over the last few years, the InterAction Council has focused on global health security in response to the Ebola and Zika outbreaks with a view to identifying policy responses that strengthen sustainable health systems and as part of a wider contribution to global security. The relevance of health to the climate change and sustainable development agendas has also been explored, with a series of expert meetings, and recently presented to the InterAction Council in Dublin. This work has led to the development of the 'Dublin Charter for One Health', endorsed by the InterAction Council in June 2017.[2]

Initiatives

The aim of the 'Dublin Charter for One Health' (see Box 11.1)[3] is to promote collaborative responsibility for the health and well-being of the planet – including human beings – in order to enable sustainable development. The Dublin Charter is a statement of the values of social justice and fairness for all; preserving the planet requires an urgent transformation of political discourse. The InterAction Council has previously recognised the universal responsibility to conserve and sustain the planet for future generations in the Universal Declaration of Human Responsibilities. The Declaration provides that 'every person should promote sustainable development all over the world in order to assure dignity, freedom, peace, security and justice for all people'.[4] It also notes that every person has 'a responsibility to protect the air, water and soil of the earth for the sake of present inhabitants and future generations'.[4] The Charter builds on this responsibility and recommends placing a duty to develop and implement policies and practices to protect planetary health on health professionals, public health practitioners, politicians, policymakers and international civil servants working across the UN and its development agencies, as well as on academics and the wider public.

Box 11.1 The 'Dublin Charter for One Health'[3]

Preamble

The Charter for One Health is a statement of the values of social justice and fairness for all. The Charter is focused on collective action to protect and promote health and wellbeing; prevent disease and disability; and foster resilience and adaption that respond to the fragility of the planet and the obligation to safeguard those aspects of the environment that are essential for human health in the Anthropocene epoch. It builds on the Universal Declaration for Human Responsibilities.

1 A right to benefit from the earth's bounty is balanced with an obligation to respect, care for, and restore the earth and its natural resources.

In the Anthropocene epoch, in exchange for the right to benefit from the earth's bounty, we have an obligation at all levels [of] society – personal, community, national, regional, global and planetary – to respect, care for and restore the earth and its natural resources. In addition, it requires radical adjustments in the way we live, work, produce, consume, generate our energy, transport ourselves and design our cities.

In the face of threats to human health and wellbeing, the sustainability of civilization, the natural and human made systems that provides support, there are obligations and responsibilities placed on all to nourish and sustain the diversity of life within which we co-exist and on which we depend.

2 Everyone has an obligation to respect life by conserving, sustaining, and making resilient the planetary and human systems essential for the health of humanity and the wellbeing of future generations.

To support these actions, there is a need to agree on a statement of values and practices based on our interconnectedness, cooperation and individual action focused on conserving, sustaining, and making resilient the planetary and human systems on which health depends.

3 Sustainable development is dependent on peaceful societies and an obligation on all to contribute to human security.

Sustainable development is dependent on an environment of peace and stability. This is an obligation for all to invest in more inclusive societies and minimize the effects of disasters to create conditions for all people to thrive. In areas of conflict and post-conflict, we must aim to make health a bridge to peace, focusing on conflict prevention.

The InterAction Council recommends that states, health professionals, public health practitioners, politicians, academics and the wider public should adopt the principles and recommendations set out in the Charter as a pathway to addressing human and planetary health. Box 11.2 provides a summary of the main actions outlined in the 'Dublin Charter for One Health'.

Box 11.2 Summary of actions in the 'Dublin Charter for One Health'

1 Strengthen multisector solutions to achieve the SDGs.
2 Take preventive approaches for peace and security, and resilience to emerging threats.
3 Mainstream *One Health* within health systems for universal health coverage.
4 Strengthen *One Health* governance mechanisms for systems reform.
5 Build Planetary Health leadership for future generations.
6 Establish an independent accountability mechanism for *One Health*.

References

1 InterAction Council. *About us*. www.interactioncouncil.org/about-us (accessed 16 July 2018).
2 InterAction Council. *Final Communique 2017*. www.interactioncouncil.org/sites/default/files/34th%20Annual%20Plenary%20Meeting%20final%20communique.pdf (accessed 16 July 2018).
3 InterAction Council. *34th Annual Plenary Meeting concludes with Dublin Charter for One Health*. www.interactioncouncil.org/publications/34th-annual-plenary-meeting (accessed 16 July 2018).
4 InterAction Council. *A universal declaration of human responsibilities*. www.interactioncouncil.org/publications/universal-declaration-human-responsibilities (accessed 16 July 2018).

For more information about the InterAction Council, please visit www.interactioncouncil.org/about-us

6 The International Veterinary Students' Association (IVSA)

A short introduction to IVSA's Standing Committee on Veterinary Education (SCoVE) One Health workshops

Lisa Buren, 2017–2018 Chair, SCoVE

Overview

The International Veterinary Students' Association (IVSA) is a not-for-profit organisation founded in 1953 and run by volunteer veterinary students. The mission of IVSA is: 'To benefit the animals and people of the world by harnessing the potential and dedication of veterinary students to promote the international application of veterinary skills, education and knowledge'.[1]

Initiatives

In 2015, IVSA's Standing Committee on Veterinary Education (SCoVE) decided to set up a workshop on *One Health* to be given by veterinary students to elementary school students. This plan originated from the thought that *One Health* is not only an essential part of veterinary education, but should be part of all education since it impacts on everyone's life in direct or indirect ways. The workshop was eventually developed in collaboration with IVSA's Standing Committee on One Health (SCOH). After this workshop was delivered, a further workshop was developed for 12–18-year-olds. The aim of both workshops was to foster an understanding of the *One Health* approach in order to improve health and well-being at the interface between humans, animals and their environments.

One Health workshop for 7–12-year-olds

The target audience was elementary school students in different countries, with various cultural backgrounds and upbringing. The aims were to:

- develop an understanding of the diversity of the veterinary profession;
- develop an understanding of the importance of biodiversity, ecocentrism and *One Health*;
- increase awareness of important zoonotic diseases.

One Health workshop for 12–18-year-olds

The target audience was middle or high school students in different countries, with various cultural backgrounds and upbringing. The aims were to:

- develop an understanding of the diversity of the veterinary profession;
- develop an understanding the importance of *One Health* through increasing awareness of zoonotic diseases, food hygiene, antimicrobial resistance and sustainable development.

Future goals

SCoVE will continue to provide these workshops to veterinary students worldwide through IVSA. These students will then share the knowledge acquired through these workshops at a local level.

Reference

1 International Veterinary Students' Association. *Mission statement*. www.ivsa.org/at-glance/mission-statement/ (accessed 16 July 2018).

For more information about SCoVE, please visit www.ivsa.org/standing-committees/scove/

7 The Network for Evaluation of One Health (NEOH)

Evaluating One Health

Barbara Haesler, Chair NEOH

Overview

Several *One Health* initiatives have been implemented around the world, such as the establishment of cross-sectoral coordination and communication and data sharing mechanisms, but no standardised methodology exists for quantitative or qualitative evaluation of *One Health* activities. As a result, policy makers have limited evidence for making decisions on a systematic, integrated approach to health that aims to break up artificial boundaries created by disciplines and sectors.

The European Union COST (European Cooperation in Science and Technology – a pan-European intergovernmental framework) funded the Network for Evaluation of One Health (NEOH). This is an open network that brings together more than 200 researchers, practitioners, decision-makers and other stakeholders from 24 COST participating countries in Europe, 12 overseas countries and international organisations with an interest in *One Health* and evaluation. Experts from economics, public health, geography, evaluation, epidemiology, nutrition, animal health and anthropology, among others, collaborate to focus on an integrated, interdisciplinary evaluation of *One Health* activities. Underpinning these efforts is the recognition of complex dynamics in health systems, involving multiple, closely interrelated species, actors and dimensions spanning health, social, ecological and economic sciences and knowledge. *One Health* as it is perceived by the NEOH community is more than simply another approach to problems – it implies a paradigm shift in our appreciation of humanity, animals, plants and their cohabitation.

Initiatives

The network is organised into four working groups that have sequential tasks of: developing a handbook for evaluation of *One Health* (WG1); applying it to case studies (WG2); conducting a meta-analysis (WG3); and engaging stakeholders and disseminating outcomes (WG4). These working groups rely on active and motivating leadership for their own goals, but also within the complete network. Furthermore, the network is growing as work progresses. NEOH aims to convey an open and tolerant attitude, welcoming new ideas, integrating as many aspects as possible and seeking expertise where it feels there is a lack. This has led to a core group of active members who contribute to the different working groups and frequently exchange information with a changing group of more peripheral experts who contribute sporadically. The concept of adaptive leadership and leading by inspiration has contributed significantly to the innovation of methods and integration of existing knowledge from multiple fields. The result is a framework primarily validated by the expertise of the contributors and tested on a dozen case studies from Europe and Africa focusing on zoonotic diseases, surveillance capacity building, obesity, animal welfare, *One Health* education and antimicrobial resistance. This provides a solid starting point to orient future evaluation research in *One Health* and the established working attitude sets the Network up for future collaborations.

The Network has generated great momentum in the exchange and integration of knowledge and experiences from various fields. Many fields of expertise are still needed in the community such as stronger representation from social sciences, legal science, philosophy, theology, psychology and education. Network participants intentionally keep the work open and unrestrained to avoid the creation of a new silo. Attempts are underway to secure funds to continue the established Network into the next phase where we can continue to promote our values and contribute to shaping the values of the wider scientific community and, ultimately, society at large. The future structure of NEOH may take the form of a Community of Practice of international society.

For more information about NEOH, please visit http://neoh.onehealthglobal.net/

8 The One Health Commission (OHC)

Connecting the Global One Health Community

Joann Lindenmayer, Board Director and Cheryl Stroud, Executive Director

Overview

The One Health Commission (OHC), a non-profit, non-governmental organisation (NGO), created in the US in 2009 as the result of a partnership between the American Veterinary Medical Association (AVMA) and the American Medical Association (AMA). The resulting 'One Health Initiative Task Force' recommended the creation of an organisation that would actively promote *One Health* and the OHC was formed. Today, the Commission works to enhance collaborations among all professionals and to increase awareness at all levels of society of the interconnectedness of people, animals and the environment. Through 'One Health Action Teams' made up of individuals from diverse backgrounds, the OHC identifies gaps and works to address them. Examples include the OHC Bat Rabies Education Team, the One Health Education Task Force (see below) and two newly formed teams, the One Health – Social Sciences Team and the One Health-Ecology/Environment Team. Teams to address One Health – Comparative Research/Medicine, One Health – Food Safety/Security and One Health – Advocacy/Public Policy and others are due to be formed.

Initiatives

What are the OHC's strategic priorities and how are they significant?

1 *Connecting stakeholders working in the One Health space: We can do more together!* The OHC has held two international 'Who's Who in One Health' webinars to meet the *One Health* community's insatiable appetite to know who is doing what for *One Health* and to connect them so they can all work together. A 'Who's Who in One Health' webpage is provided and a Global One Health Community listserv that shares information in a 'One Health Happenings' newsletter. The long-term goal is: to facilitate interactions between organisations working in parallel paths, each with its unique strength, towards a common end point of human, animal and planetary health and well-being; and to form a synergistic, global (*One Health*) Alliance. Towards that end, in 2016 the OHC partnered with the One Health Platform and the One Health Initiative to create an annual, global One Health Day celebrated each year on 3 November. The inaugural One Health Day saw more than 155 global events in over 35 countries.
2 *Educating about One Health and One Health Issues.* The OHC identifies gaps and issues that need more attention and hosts global educational webinars. It also provides its online webinar platform for unincorporated *One Health*

groups that do not otherwise have a means to share their *One Health* educational actions with a larger global audience.

3 *Nurturing the Next Generation of One Health leaders*. This priority takes two forms.

- *Primary/Secondary Education*. The Commission believes very strongly that humanity will not overcome many of its serious global problems without reframing our attitudes about our human place in the ecological continuum on Earth. A One Health Education Task Force (OHETF) was formed in late 2015 to discover how to educate about *One Health* and *One Health* issues starting very early in children's lives and progressing through adulthood. In 2016, the OHC developed an ongoing initiative, 'Preparing society to create the world we need through One Health education', and is actively seeking funding to develop teacher workshops and team-developed *One Health* curricula. This is another reason why the OHC advocates the involvement of social scientists so strongly – we'll never get anywhere without them.
- *Nurturing undergraduate, graduate and professional students working for One Health*. In 2015, the OHC brought student representatives from animal, environmental/ecology and human health domains onto its Board of Directors as full voting Board members (a position is being held for a student from the plant domain when such a student emerges). The OHC provides these students with an online meeting platform, webpages, a student listserv and encouragement for the development of student-led *One Health* projects with the intention of facilitating the formation of relationships across professions very early in their careers. These student representatives have now formed an independent Students for One Health (SOH) organisation for which the OHC will serve as a parent organisation. The first SOH Executive Council has recently been selected by students. These are our future global leaders.

For more information about the OHC, please visit www.onehealthcommission.org/

9 The One Health Initiative (OHI)

'Challenges and opportunities for the future of One Health'[1]

Bruce Kaplan, Laura Kahn, OHI Co-Founders; Thomas Monath, Consultant to Biotechnology Industry; Lisa Conti, Deputy Commissioner and Chief Science Officer of the Florida Department of Agriculture and Consumer Services

Overview

The One Health Initiative (OHI) autonomous pro bono team was originally established in 2006–2007. The OHI website, created in October 2008, has been described as 'the international clearing house for significant One Health

Box 11.3 One Health Initiative's (OHI's) mission and vision

Mission statement

Recognizing that human health (including mental health via the human–animal bond phenomenon), animal health, and ecosystem health are inextricably linked, One Health seeks to promote, improve, and defend the health and well-being of all species by enhancing cooperation and collaboration between physicians, veterinarians, other scientific health and environmental professionals and by promoting strengths in leadership and management to achieve these goals.

Vision statement

One Health (formerly called One Medicine) is dedicated to improving the lives of all species – human and animal – through the integration of human medicine, veterinary medicine and environmental science.

One Health shall be achieved through:

1 Joint educational efforts between human medical, veterinary medical schools, and schools of public health and the environment;
2 Joint communication efforts in journals, at conferences, and via allied health networks;
3 Joint efforts in clinical care through the assessment, treatment and prevention of cross-species disease transmission;
4 Joint cross-species disease surveillance and control efforts in public health;
5 Joint efforts in better understanding of cross-species disease transmission through comparative medicine and environmental research;
6 Joint efforts in the development and evaluation of new diagnostic methods, medicines and vaccines for the prevention and control of diseases across species and;
7 Joint efforts to inform and educate political leaders and the public sector through accurate media publications.[2]

information' and, by some, the 'New York Times of One Health'. OHI's world-wide circulation has been strengthened by its up-to-date *One Health News*, publications and upcoming events postings relative to animal, human and environmental health. The OHI website is currently estimated to be accessed monthly by roughly 20,000 individual visitors from over 150 countries; it is known to have had at least 119 reciprocal links worldwide. Among leading high-profile search engines, the OHI website is listed first or among the first when either 'One Health Initiative' or 'One Health' are searched.

Initiatives

OHI considers it critically important to establish an institutionalised international *One Health* approach now and in future.[1] Co-equal transdisciplinary collaborations are essential for efficacious and expeditious global public health and comparative medicine research advancements. Abundant and verifiable corroborative scientific examples are documented on *One Health* websites and in professional journals, textbooks and newsletters from the nineteenth, twentieth and twenty-first centuries. These include issues such as:

> Heart Disease, Cancer, Orthopedic Disease, Anesthesiology, Obesity, Parasitic Diseases, Tuberculosis, Global Infectious Disease, Influenza, Human Hepatitis C virus, Tickborne Diseases, Food Safety, Hendra virus vaccine, Aspergillus felis, Immunizations (vaccinations), Lou Gehrig's Disease, Ebola, Antibiotic Use and Resistance, Staphylococcus resistant infections, Environmental health Policymaking, Crimean-Congo hemorrhagic fever viruses, Renderpest, emergency/Disaster preparedness and many others.[3]

References

1 Kaplan B, Kahn LH, Conti LA, Monath TP, Yuill TM. Challenges and opportunities. *Pan European Networks, Science & Technology*, December 2017: 166–169. http://edition.pagesuite-professional.co.uk/html5/reader/production/default.aspx?pnum=167&edid=c780812e-ef43-4ea4-bf5a-7c8f0cfe1e7d&isshared=true (accessed 16 July 2018).

2 One Health Initiative. *Mission statement.* http://onehealthinitiative.com/mission.php (accessed 16 July 2018).

3 Kahn LH, Kaplan B, Monath TP, Woodall J, Conti LA. *'One Health', a significant international public health/comparative medicine paradigm shift! Why? How?* www.onehealthinitiative.com/publications/ONE%20HEALTH%20a%20significant%20international%20public%20health%20comparative%20medicine%20OHI%20POSTING%20May%2013%202015.pdf (accessed 16 July 2018).

For more information about the OHI, please visit www.onehealthinitiative.com

10 The One Health Platform (OHP)

Enhancing the understanding of and preparedness for the current and future outbreaks of disease

Chair, Ab Osterhaus, *Professor of Wildlife Virology and Virus Discovery; Vice-chair*, John McKenzie, *Research Associate and Professor of Tropical Infectious Diseases at Curtin University, Australia; Secretary*, Chris Vanlangendonck, *Communication Strategist at Semiotics, Belgium*

Overview

Mission

The One Health Platform (OHP) is a strategic forum of stakeholders and a *One Health* reference network that aims to enhance understanding of and preparedness for current and future outbreaks of zoonoses, emerging infectious diseases in humans and animals and antimicrobial resistance – including the ecological and environmental factors which impact on these diseases.

Need for a One Health Platform

Recent global disease events have underpinned the increasing effects of zoonotic diseases on human and animal health. It has also become clear that changes in the environment, such as population growth and climate change, are drivers for the emergence of such zoonoses. The initiators of OHP have therefore set up an independent organisation that promotes an integrated approach to combatting different types of health threats: (re-)emerging neglected viruses, antiviral and antibiotic resistance and bacterial and parasitic infections. By OHP, they wish to forge a community of stakeholders and to stimulate information sharing and interaction at all levels.

Initiatives

Objectives

OHP aims to:

1 provide a strategic forum for researchers, early career investigators, governmental and non-governmental institutions, international organisations and companies to foster cross-sectoral collaborations;
2 identify and prioritise research gaps in the fields of zoonoses, emerging infectious diseases and antimicrobial resistance, including the ecological and environmental factors which impact on these diseases, and advocate the resulting scientific research agenda – both on a scientific and a policy level;

3 create synergies and facilitate the sharing of data between researchers and research groups in order to mend the prioritised research gaps and translate these novel data to anyone who might benefit from them;
4 disseminate the results and insights of existing and new research projects on zoonoses, emerging infectious diseases and antimicrobial resistance, including the ecological and environmental factors which impact on these diseases;
5 establish an information reference centre for the *One Health* community;
6 enhance awareness of the value of the *One health* approach through communication; facilitation of interactions between stakeholder groups; education and training; and specific efforts to convince 'non-believers', both in the research community and in the policy arena.

International One Health Coalition

The International One Health Coalition is a collaborative partnership between OHP and existing international governmental and non-governmental organisations and institutions. With the aim of reinforcing the *One Health* concept, the coalition partners set up a framework for information sharing, cooperation and awareness- raising activities. Through the International One Health Coalition, the One Health Platform Foundation builds a true global community of *One Health* stakeholders.

For more information about the OHP, please visit www.onehealthplatform.com

11 The One Health Tripartite collaboration (FAO,[1] OIE[2] and WHO[3])

Sharing responsibilities and coordinating global activities

Tianna Brand, Julie Sinclair and Matthew Stone

Overview

The Tripartite has been working together on issues of mutual interest for almost a quarter of a century, but it was only in 2010 that they formally outlined their collaboration in a *One Health* concept note, 'Sharing responsibilities and coordinating global activities to address health risks at the human–animal ecosystems interfaces'.[4]

Since that time, this document has served as the basis for how the international organisations collaborate together. This collaboration was subsequently put into action, leading to the 'High-level technical meeting to address health risks at the human–animal–ecosystems interfaces'[5] in Mexico in 2011, organised by the Tripartite along with UN System Influenza Coordination. Three specific topics identified were zoonotic influenza, rabies, and antimicrobial resistance (AMR) – high priority representative examples where cross-sectoral collaboration could offer increased efficiency and effectiveness.

Initiatives

Addressing the priority topics cited above led to the development of a global action plan against AMR; a global framework for the elimination of dog-mediated human rabies; and enhancement of zoonotic influenza global surveillance and information sharing through the global network of expertise on animal influenza (OFFLU) network and with WHO experts to adapt flu vaccines to the current animal strains in circulation. Through this collaboration, the Tripartite developed a set of standard best practices for naming new human infectious diseases, with the aim of minimising unnecessary negative effects on nations, economies, people and animals.[2] Other topics that have surfaced as priorities for collaboration that have therefore also been addressed by the Tripartite include zoonotic tuberculosis, Middle East respiratory syndrome-related coronavirus (MERS-CoV), strengthening of cooperation between human health and animal health systems and the synergistic capacity development of health services.

The scope and breadth of the collaboration continues to thrive. Building on the momentum of these three topics, the Tripartite published a document in 2017 entitled, 'The tripartite's commitment – providing multi-sectoral, collaborative leadership in addressing health challenges'.[6]

Current strategic priorities, as outlined in the document mentioned above, include:

- supporting the Interagency Coordination Group for AMR established through a declaration adopted by the UN General Assembly;
- designing and implementing national health services' capacity development programmes as a result of evaluations and assessments;
- strengthening and modernisation of early warning and surveillance/monitoring systems;
- improving collaboration in foresight, preparedness and response to emerging, re-emerging and neglected infectious diseases;
- using a multisector approach to addressing food safety challenges; and,
- encouraging and promoting coordinated research and development to achieve a common understanding of the highest priority zoonotic diseases.

The broadening of the Tripartite's agenda is in recognition of the increasing complexity of issues that require a multisectoral approach, and the expectation of the organisations' members that the Tripartite will continue its leadership role in setting the example for multisector collaboration.[7]

References

1 FAO (Food and Agriculture Organization of the United Nations). www.fao.org/about/en/ (accessed 16 July 2018).
2 OIE (World Organisation for Animal Health). www.oie.int/about-us/ (accessed 16 July 2018).
3 WHO (World Health Organization). www.who.int/about/en/ (accessed 16 July 2018).
4 The FAO–OIE–WHO collaboration, tripartite concept note. *Sharing responsibilities and coordinating global activities to address health risks at the animal–human–ecosystems interfaces*. www.oie.int/fileadmin/Home/eng/Current_Scientific_Issues/docs/pdf/FINAL_CONCEPT_NOTE_Hanoi.pdf (accessed 16 July 2018).
5 High-level technical meeting to address health risks at the human–animal–ecosystems interfaces, Mexico City, 15–17 November 2011. Food and Agriculture Organization of the United Nations, World Organisation for Animal Health and World Health Organization. http://apps.who.int/iris/bitstream/handle/10665/78100/9789241504676_eng.pdf;jsessionid=480E071E7FE3F16363C426C0D15AD34A?sequence=1 (accessed 2 July 2018).
6 *The tripartite's commitment – Providing multi-sectoral, collaborative leadership in addressing health challenges*. Food and Agriculture Organization of the United Nations, World Organisation for Animal Health and World Health Organization. www.oie.int/file admin/Home/eng/Media_Center/docs/pdf/Tripartite_2017.pdf (accessed 2 July 2018).
7 Tripartite partnership of FAO, WHO, and OIE highlights the importance for strengthened work at the human–animal–ecosystem interface. *Food and Agriculture Organization of the United Nations* [news archive], 25 February 2013. www.fao.org/ag/againfo/home/en/news_archive/AGA_in_action/2013_Tripartite_partnership_at_the_human-animal-ecosystem_interface.html (accessed 2 July 2018).

For more information about the One Health Tripartite, please visit www.oie.int/

12 The Planetary Health Alliance (PHA)

Optimising human health and environmental stewardship objectives

Sam Myers, Chris Golden, Steven Osofsky and Amalia Almada,
Senior Programme Manager

Overview

The Planetary Health Alliance (PHA) is a consortium of over 95 dedicated universities, NGOs, research institutes and other partners from over 26 countries with a shared mission – supporting the growth of a rigorous, policy-focused, transdisciplinary field aimed at understanding and addressing the human health implications of accelerating environmental change.

Initiatives

The PHA forges collaborations across a broad array of scientific disciplines, as well as with decision-makers, movement builders and local communities, to build a rigorous evidence base for informing policy solutions aimed at optimising both human health and environmental stewardship objectives. The consortium works to support the growth of the planetary health field by:

- connecting the global planetary health community via an annual meeting and virtual forums;
- facilitating the training of the next generation of scientists through publicly accessible educational resources and a postdoctoral fellowship;
- communicating planetary health science for education and policy applications; and
- applying planetary health science directly to inform policy and practice.

For more information about the PHA, please visit https://planetaryhealthalliance.org/

13 The Southeast Asia One Health University Network (SEAOHUN)

Developing a well-prepared One Health workforce to prevent, detect and respond to the threat of infectious diseases

Vipat Kuruchittham, Executive Director

Overview

The spillover of infectious agents from animals to humans in recent decades has had a significant impact on the health of humans, animals, our environment and the global economy. To minimise the impact of future pandemic threats, the Southeast Asia One Health University Network (SEAOHUN) was established in 2011, with support from the United States Agency for International Development's Emerging Pandemic Threats programme, to develop a well-prepared *One Health* workforce to prevent, detect and respond to the threat of infectious diseases.

Since its establishment, SEAOHUN has expanded its network membership to include 62 universities and 115 faculties, as well as government agencies, in Indonesia, Malaysia, Thailand and Vietnam. Members come from multiple disciplines connected to the health of humans, animals and the environment. SEAOHUN's educational programmes promote and facilitate collaborative activities and projects among the four countries, as well as Cambodia, Lao PDR, and Myanmar, with an emphasis on multisectoral, trans-disciplinary and trans-boundary partnerships. For its programmes and activities, SEAOHUN has partnered with public and private education institutions, national governments, inter-governmental organisations, regional organisations and international NGOs, to prepare today's workforce to combat tomorrow's infectious diseases.

Initiatives

Examples of SEAOHUN's educational initiatives include: development of *One Health* core and technical competencies; curriculum development and strengthening; introducing innovative teaching methodologies such as *One Health* problem-based learning case studies; building research capacity to provide an evidence-base for *One Health* advocacy; student and faculty exchanges; student clubs; student field attachments; and scholarship and fellowship programmes. Up until 2017, over 6,000 students, health professionals, and faculty members attended trainings, and 24 *One Health* fellows and 18 scholars were supported. Additionally, 49 community outreach activities and 37 multisectoral and multilateral events were coordinated, and 15 evidence-based materials were created.

SEAOHUN is strengthening its organisational capacity by mobilising additional resources to expand country membership, programmes and activities to engage all ten members of the Association of Southeast Asian Nations (ASEAN). SEAOHUN aims to become a leading regional organisation building the skills, knowledge and attitudes of the current and future *One Health* workforce, generating and sharing up-to-date knowledge on *One Health* and providing support for *One Health* policy advocacy.

For more information about SEAOHUN, please visit http://seaohun.org/

14 The United Nations Development Programme (UNDP)

Douglas Webb, Brian Lutz and Natalia Linou

Overview

The SDGs came into effect in January 2016, and they will continue to guide UNDP policy and funding until 2030. As the lead UN development agency, UNDP is uniquely placed to help implement the Goals through our work in some 170 countries and territories.

UNDP's strategic plan focuses on key areas including poverty alleviation; democratic governance and peacebuilding; climate change and disaster risk; and economic inequality. UNDP provides support to governments to integrate the SDGs into their national development plans and policies.[1] This work is already underway, as UNDP supports many countries in accelerating progress already achieved under the Millennium Development Goals. UNDP's track record working across multiple goals provides the organisation with valuable experience and proven policy expertise to ensure the targets set out in the SDGs are reached by 2030. But it cannot do this alone.

Achieving the SDGs requires the partnership of governments, private sector, civil society and citizens alike to make sure we leave a better planet for future generations.

Initiatives

The links below provide more information about the areas UNDP is focused upon:

- *Sustainable development* – www.undp.org/content/undp/en/home/sustainable-development.html.
- *Support for implementation of the SDGs* –www.undp.org/content/undp/en/home/librarypage/sustainable-development-goals/undp-support-to-the-implementation-of-the-2030-agenda.html.
- *Support for country implementation of the SDGs* – www.undp.org/content/undp/en/home/librarypage/sustainable-development-goals/the-sustainable-development-goals-are-coming-to-life.html.
- *Financing the SDGs in the least developed countries* –www.undp.org/content/undp/en/home/librarypage/poverty-reduction/afd-undp-financing-sustainable-development-and-managing-vulnerab.html.
- *Democratic governance and peacebuilding* – www.undp.org/content/undp/en/home/democratic-governance-and-peacebuilding.html.
- *Climate and disaster resilience* –www.undp.org/content/undp/en/home/climate-and-disaster-resilience.html.
- *Gender equality* – www.undp.org/content/undp/en/home/gender-equality.html.

- *Crisis response* – www.undp.org/content/undp/en/home/crisis-response.html.
- *Development impact* – www.undp.org/content/undp/en/home/development-impact.html.

UNDP's support for country implementation of the 2030 Agenda and the SDGs is framed by the common approach of the UN Development Group called MAPS[1] (Mainstreaming, Acceleration and Policy Support).[2]

References

1 UNDP. *UNDP projects around the world.* http://open.undp.org/#2018 (accessed 16 July 2018).
2 UNDP. *About the UNDP.* www.undp.org/content/undp/en/home/sustainable-development-goals/resources.html (accessed 16 July 2018).

For more information about UNDP, please visit www.undp.org/content/undp/en/home/sustainable-development-goals/resources.html

15 The World Bank Group (WBG)

World banking on One Health

Timothy Bouley, Climate Change and Health Programme Lead

Overview

The World Bank[1] has for many years, and from multiple sectors, been building and investing in *One Health*-related systems. In 2018, this work culminated in the release of the World Bank 'Operational framework for strengthening human, animal, and environmental public health systems at their interface' – a systematic *One Health* focused effort to establish better investments that prevent, prepare and respond to diseases and antimicrobial resistance issues.[2] This work built on previous World Bank studies over the past decade that have explored international financial institution entry points for *One Health* work, such as: 'People, pathogens and our planet', Volumes I and II; 'Connecting sectors and systems for health results'; 'Reducing climate-sensitive disease risks'; 'Climate-smart healthcare: Low carbon and resilience strategies for the health sector'; and 'Drug-resistant infections: A threat to our economic future'. Dozens of white papers, speeches and notes have also addressed the issue with a refined regional and country level focus.

Initiatives

In collaboration with other global organisations, such as WHO and OIE, the World Bank has supported the development of an environmental health bridging framework to complement tools already in practice that are increasing uptake of international health regulations and veterinary services. In terms of direct investments, the World Bank has made many that serve overall *One Health* aims. For example, the institution has supported developing countries to implement the first global public health programme for avian influenza control and human pandemic preparedness and response (i.e. 'Global Programme for Avian Influenza') and deployed the Regional Disease Surveillance System Enhancement programme in West Africa, a several hundred million dollar investment towards reducing disease outbreaks across human and animal health.

Investing in One Health

The economic argument for investing in *One Health* is compelling. The West African Ebola outbreak, for example, resulted in a loss of 12 per cent combined GDP growth in Guinea, Liberia and Sierra Leone. Zoonoses frequently account for acute and persistent disruptions in education, routine health care and livelihood generation. While economic impacts are considerable, investments in capacity to mitigate risk are still very limited. *One Health* strategies are very cost effective, not only for reducing pandemic and antimicrobial resistance risks, but also endemic diseases. Annual investment of approximately US$1.9–3.4 billion to raise human and animal health system capacity in World Bank client countries is expected to return upwards of US$30 billion per year in avoided losses. This is a high return on investment even if only a portion of pandemics are prevented, while also generating gains for agricultural production and control of endemic diseases.

Rationales for One Health engagement

The World Bank's rationale for involvement in *One Health* work is manifold. The following points provide some of the justification.

- The World Bank has both a global reach and engages in all the sectors involved (public health, animal health, environment, disaster risk management, global risk communications). Few development institutions combine such a country-level track record of engagement through lending and economic work in all these sectors – and, moreover, a capacity for global scope in delivery.
- The World Bank has valuable operational expertise supporting multisectoral programmes, from design to appraisal to implementation of substantial investments and related policies, working to improve coherence and coordination across sectors.
- The World Bank can finance and mobilise additional resources for these programmes; the projected incremental annual costs of human–animal–environment public health systems are well above the capacity of UN agencies to manage.
- The World Bank has a mandate to work on the provision of global public goods (GPGs) (prevention of infectious diseases is a grossly undersupplied GPG and the only health issue that meets the World Bank Development Committee's narrow definition of GPGs). The World Bank was a critical donor in mobilising funds for Ebola and highly pathogenic avian influenza (HPAI) response, and thus has a strong incentive to invest in cost-effective multi-disease prevention and preparedness capacity to avoid high ad hoc emergency response costs.
- Tackling threats such as zoonoses and antimicrobial resistance is a pressing and increasingly severe development challenge, with significant impacts on health, poverty, food security, nutrition, trade, environmental outcomes, ecosystem dynamics and food safety in poor countries. Addressing these present and future health risks is fully in line with the World Bank's mandate and International Development Association (IDA) agreements.
- The World Bank has extensive experience with emergency responses and with operations in fragile states. This experience offers lessons for the task of strengthening systems for disease prevention and preparedness for control actions that can be valuable for partner institutions.
- The World Bank played a leading role in the response to avian influenza and Ebola, and thus has a stock of experience and established relationships within countries and among other stakeholders, as well as credibility.

- The World Bank is committed to increasing cross-pollination of interests and collaboration among teams. The recent World Bank public health policy document, 'Connecting sectors and systems for health results', sets out a multi-pillared approach to achieving health goals and emphasises galvanisation of actors outside the traditional public health sphere.
- The World Bank has developed privileged relationships with key international actors such as WHO, OIE, FAO, Convention on Biological Diversity (CBD), UNEP and United Nations International Strategy for Disaster Reduction (UNISDR) that can support global partnerships and enable implementation synergies in client countries.

There is no one-size-fits-all approach for *One Health* implementation. Yet this flexibility means there is ample opportunity for new global implementation across new levels and innovative programmatic objectives that draw upon *One*

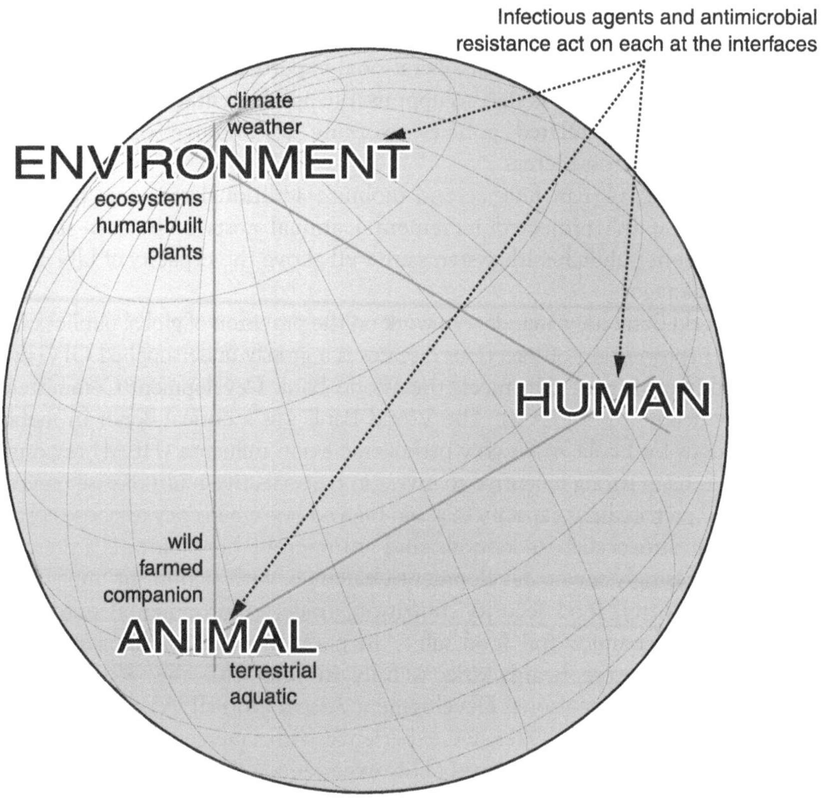

Figure 11.1 Infectious disease and antimicrobial resistance are primary areas of focus of the World Bank *One Health* operational framework.

Source: The World Bank. *Operational framework for strengthening human, animal, and environmental public health systems at their interface* (2018).

Health principles. The World Bank is committed to doing so, enhancing its already broad portfolio of work with the ultimate aim of turning *One Health* investment into *One Health* action and success.

References

1 The World Bank. *About the World Bank*. www.worldbank.org/en/about (accessed 16 July 2018).

2 The World Bank. *Operational framework for strengthening human, animal, and environmental public health systems at their interface*. http://documents.worldbank.org/curated/en/703711517234402168/pdf/WP-P133691-PUBLIC.pdf (accessed 2 July 2018).

For more information about the WBG, please visit www.worldbank.org/

16 The World Veterinary Association (WVA) and World Medical Association (WMA) One Health collaboration

WVA and WMA – a global example of One Health in action

Zeev Noga, Veterinary Policy Officer (WVA) and Otmar Kloiber, Secretary General (WMA)

Overview

The World Veterinary Association (WVA)[1] and World Medical Association (WMA)[2] collaboration agreement highlights the importance of the *One Health* concept as the integrative effort of multiple disciplines working together to attain optimal health for people, animals and the environment. In October 2012, the WVA, representing the global veterinary profession and the WMA, representing the global medical profession, signed a Memorandum of Understanding to collaborate in a unified approach to tackling common health issues and improving global health.

Initiatives

The WVA–WMA collaboration plan focuses on zoonotic diseases, responsible use of antimicrobials and enhancing collaboration on education, clinical care and public health. WVA and WMA hold regular meetings to discuss issues of common interest and participate in the each other's events to highlight the work and progress of their professions which address health concerns on a global level.

In recent years, the two organisations have prepared and published joint statements and supported each other on various health issues such as rabies control, antibiotic resistance and the regular availability of ketamine (non-scheduling) as well as on the *One Health* concept.

In May 2015, for the first time in history, the two world associations organised a global conference on *One Health* with the theme, 'Drivers towards One Health – Strengthening collaboration between physicians and veterinarians'.[3] The event signified an acknowledgement by both disciplines, at the highest levels, of the importance of cooperation between the veterinary and medical professions. The conference concluded that in order to make a difference, the health professions need to work on leadership, networking, cooperation, communication, facilitation and building trust at all levels, starting with veterinary and medical schools.

Following this successful conference, the WVA and WMA, in close collaboration with the Japanese Medical Association (JMA) and the Japanese Veterinary Medical Association (JVMA), organised a second conference with the theme, 'Moving forward from One Health concept to One Health approach',[4] which took place in November 2016 in Japan. The main outcome of this conference was the historic signature on the Memorandum of Fukuoka[5,6] by WVA, WMA, JMA and JVMA.

References

1 World Veterinary Association. *About WVA*. www.worldvet.org/about.php (accessed 16 July 2018).

2 World Medical Association. *Who we are*. www.wma.net/who-we-are/ (accessed 16 July 2018).

3 World Veterinary Association–World Medical Association. *Drivers towards One Health – Strengthening collaboration between medical and animal doctors*. www.worldvet.org/events.php?item=89&view=item (accessed 16 July 2018).

4 World Veterinary Association–World Medical Association and Japan Medical Association–Japan Veterinary Association. *Moving forward from One Health concept to One Health approach*, 10–11 November 2016, Kitakyushu City, Fukuoka, Japan. www.ncbi.nlm.nih.gov/pmc/articles/PMC5476991/ (accessed 16 July 2018).

5 World Veterinary Association–World Medical Association. *Memorandum of Fukuoka*. www.cimm-icmm.org/page/anglais/Rapport2ndGlobalConferenceJapan2016ANG.php (accessed 16 July 2018).

6 Lueddeke GR. The UN-2030 Sustainable Development Goals and the One Health concept: A case for synergistic collaboration toward a common cause. *World Medical Association Journal* 2016; 62(4): 162–167. https://lab.arstubiedriba.lv/WMJ/vol62/december-2016/#page=44 (accessed 16 July 2018).

For more information about the WVA–WMA One Health collaboration, please visit www.wma.net/wp-content/uploads/2017/01/2GCOH-Japan-SUMMARY-2016.pdf

17 The World Wildlife Fund (WWF)

Resetting the Global Agenda in 2020

Martin O'Halloran, Editorial Manager, WWF International

Overview

The World Wildlife Fund for Nature (WWF) is one of the world's largest and most experienced independent conservation organisations, with over 30 million followers and a global network active in nearly 100 countries. Our mission is to stop the degradation of the planet's natural environment and to build a future in which people live in harmony with nature, by conserving the world's biological diversity, ensuring that the use of renewable natural resources is sustainable and promoting the reduction of pollution and wasteful consumption.

The challenge

WWF has been here for more than half a century, and during that time the world has changed.

The impact of many success stories has been to save species and natural places and drive our economy and society towards a more sustainable path. But wildlife numbers globally have fallen by more than one-half. We continue to see forests destroyed, coral reefs dying, rivers running dry and species becoming extinct at an alarming rate. Climate change is starting to bite. And, as a growing human population consumes ever more, the pressures on nature continue to mount.

If the loss of nature is not reversed, the natural systems of our planet will be seriously damaged and our civilisation will head into volatile, dangerous territory as this century progresses. The future of humankind and much of life on Earth, as we know it, is at risk.

Turning this around is the biggest task humanity has ever faced. And what happens over the next few years will determine our future.

Our strategy

WWF has a unique responsibility and pivotal role in scaling up global conservation efforts in the coming decade. The organisation is determined to make its contribution count, capitalising on its strengths: the local and global reach of its network; the global influence of its brand; its work on international systems as well as in the field; and its huge public support across the whole world.

WWF has agreed a set of seven integrated outcomes that the WWF network will work with many partners to achieve by 2030. They are ambitious and challenging, but necessary.

WWF will work – from local people-led conservation to influencing the biggest global drivers – towards achieving the following global outcomes by 2030:

- protect and restore at least 30 per cent of land and sea
- eliminate the illegal wildlife trade and reduce overexploitation
- halt deforestation and degradation
- keep the most important rivers flowing
- double sustainable fisheries
- halve the impacts of the food system
- reduce greenhouse gas emissions by 40 per cent.

Driven by a greater ambition WWF is uniting its network in identifying areas where they will collectively achieve greater collective impact. They are boosting learning and innovation, developing new 'high impact initiatives' where they see the urgency and opportunity to lead rapid progress and, in everything they do, looking for powerful partnerships and coalitions able to design solutions that deliver the new ambition.

The high impact initiatives WWF is currently discussing with partners include: shutting down Asian ivory markets and lowering demand for ivory products; doubling tiger numbers in their heartlands; mobilising finance at scale for securing large protected area networks; rescuing coral reefs; encouraging conservation led by communities and indigenous peoples; and introducing reliable and impactful green bonds, among others.

Initiatives

We are rapidly approaching an unmissable opportunity for the world to change direction.

The global response to the accelerating loss of nature is slowly gathering momentum. UN initiatives such as the SDGs, the Paris Climate Agreement and the Convention on Biological Diversity (CBD) point the way forward, with a series of targets set for 2020. Although this is exciting progress, the current commitments are not enough. These global initiatives need to be co-ordinated, unified and deepened over the coming decade.

The year 2020 will be a historic moment when world leaders will take key decisions on the environment, climate and sustainable development. These decisions will set the agenda for the next decade. We must influence these global decisions and send a message that it is no longer acceptable to continue destroying our natural world – and that we can take a different, better path. Put simply, we need governments to commit by 2020 to reversing the loss of nature and living within our environmental means. Only a new determination for greater impact can put us on track to achieving the 2030 targets.

WWF's combined advocacy and communications initiatives will support efforts to reset the global agenda by 2020. For example, WWF is partnering with the CBD to deliver one of their key 2020 targets: helping people around the world understand the importance of biodiversity and how they can conserve it and use it sustainably. And as a partner of the forthcoming Netflix natural history series 'Our planet' (to be screened in 2019), WWF will get the opportunity to reach out to millions of viewers.

The importance of the coming two years cannot be overstated. WWF fully intends to play its part and calls on others to do the same.

To for more information about WWF, please visit http://wwf.panda.org/

12 Leading in an era of uncertainty, upheaval and anxiety

> As long as poverty, injustice, and gross inequality persist in our world, none of us can truly rest.
>
> (Nelson Mandela, former president of South Africa)

Imagining a better world: towards a new World Order?

In an article entitled, 'We can plug Silicon Valley's moral void',[1] Rohan Silva, former policy advisor to then UK prime minister David Cameron, draws on philosopher Karl Popper's *The Open Society and Its Enemies*, published in 1945.[2] Claiming that the book is 'one of the most important works in the history of western philosophy',[1] Silva draws attention to Popper's main theme that 'the root cause of flawed ideologies' is 'utopianism – the idea that human nature can be perfected or that an ideal society can be constructed'.

Silva highlights that 'utopian ideologies are still with us today', evidenced in Silicon Valley 'where the technology industry has been deeply shaped by the libertarian movements of the 1960s and 1970s'.[1] The worldview that individuals 'should always and everywhere be free from any authority or top-down control – especially from government', found its way, Silva observes, in the 'democratisation of information, giving rise to search engines and blogs as well as social media' that deliver 'vast benefits' but also downsides – 'extremist materials … hate crimes and bullying'.[1] Silva's advice for policy-makers is to start by 'recognising the dangers of utopianism, challenging libertarians and adopting British values' that are 'sceptical and pragmatic rather than utopian and fundamentalist', introducing 'tough laws' to counter 'terrorist content' and 'tackle cyberbullying with severe punishment for firms that refuse to comply' and putting in place 'strict controls to protect children'.[1]

Silva may agree with Jenni Russel's review of Steven Pinker's book, *Enlightenment Now*,[3] which 'argues that contemporary gloom and fear is quite misplaced' and all that is needed is the belief that 'the appeal of liberal democracies and the enlightenment values that underpin them are so powerful they need only to be spelt out to be accepted' – because 'reason makes us recognise the humanity of others'.[4] Although well-intentioned, some of Pinker's assumptions may benefit from further investigation, especially when considering previous 'flawed ideologies, fascism, communism and so on – that are causing vast damage around the world'.[4] We might add to this Alex Massie's observations that the US 'is pivoting to the Pacific when it isn't busy turning in on itself and ripping up the nuke books that govern international affairs' while 'Europe also looks nervously eastwards and doesn't much care for what it sees' and about a third of European nations 'think representative democracy' is 'a bad way of governing their countries' and tend towards 'autocracy'.[5] Indeed, while some would prefer 'a reformed capitalism … a renewed spirit of citizenship and democratic engagement', Massie returns to the 'nagging question "how?"'.[5]

While it is true, as Pinker asserts, that some things in the world are getting better (e.g. poverty, child mortality), it is also the case that others are staying about the same (e.g. HIV) or getting worse (e.g. regional armed conflicts, malnutrition, urban slum dwelling, mental health), evidenced by real-time Worldodometers statistical data.[6] Samuel Moyn, professor of law and professor of history at Yale University raises similar concerns and cautions 'that even the best trends are subject to frightening interruption or outright reversal'.[7] His questions, 'How much does it matter, for instance, that growth has skyrocketed, if the one percent captures so substantial a part of the gains?' and 'How will a much-hyped modernity look 100 or 1,000 years from now, on a dying or dead planet?', are reminders of the planet's vulnerabilities and of reaching overly optimistic conclusions.

Perhaps, as UK Astronomer Royal Lord Rees, asserts, one of our greatest delusions is that from a cosmic point of view, many 'still think we humans are necessarily the culmination of the evolutionary tree' when 'we are probably still nearer the beginning than the end' with about 6 billion years to go before the sun's 'fuel runs out'.[8] We are, however, as highlighted throughout this book, at a turning point – 'the anthropocene era when one species, ours, can determine the planet's fate' and where our 'collective "footprint" on the Earth is heavier than ever' and a time when 'today's decisions on environment and energy resonate centuries ahead', determining 'the fate of the entire biosphere and how future generations live'.[8] In contrast, and potentially to our peril, Lord Rees reflects that 'our planning horizons have shrunk because our lives are changing so fast' and 'the political focus is on the urgent and immediate, and the next election'.[8]

Lord Rees's emphasis that humankind can 'make or break' the destiny of the planet coupled with short-term solutions to increasingly complex problems – that often advantage only vested interests – can result in a toxic combination that could smother life as we know it. Much better, as Lord Rees remarks, would be to leave a legacy this century of 'a world that is sustainable rather than devastated'.[8]

What would also be much better would be getting behind a 'global unity of purpose', as advocated in Chapter 9. More specifically, as summarised in the UN 'World economic and social survey 2013',

> Sustainable development will need to be inclusive and take special care of the needs of the poorest and most vulnerable. Strategies need to be ambitious, action-oriented and collaborative, and to adapt to different levels of development. They will need to systemically change consumption and production patterns, and might entail, inter alia, significant price corrections; encourage the preservation of natural endowments; reduce inequality; and strengthen economic governance.[9]

It may be telling that 'the three important cross-sectoral issues: sustainable cities, food security and energy transformation'[9] remain as unmet today as they were in the years leading up to the UN-2030 SDGs and continue to demand 'better understanding of the linkages among these trends and the associated changes in economic, social and environmental conditions'.[9]

According to the authors of an article entitled, 'A vision for human well-being: Transition to social sustainability', to do so will 'require a global focus on growth in well-being instead of consumption', while acknowledging that 'communities currently in poverty will need additional consumption in order to do well'.[10] Social sustainability will also necessitate

> more research to better understand the key drivers of well-being: What do humans really need in order to live well, both physically and emotionally, and to feel satisfied with their lives? What kind of societal drivers are shaping and influencing them?[10]

Viewed from a sociopolitical perspective, these inquiries could lead, for example, to former US secretary of state, Henry Kissinger's widely acclaimed book, *World Order*.[11] Most notably, Kissinger concludes that 'the Peace of Westphalia in 1648 which ended the Thirty Years War between Catholics and Protestants with an agreement that states would no longer interfere in each other's internal affairs' was the turning point in establishing the World Order we have today, underpinned by the 'balance of power as a system for managing relations between states'.[11] In consequence, according to Kissinger's historic analysis dating back to 448 CE, the Westphalia system led to 'the search for one universal system as in Imperial China or in early Islam', while 'Europe developed a pluralistic system of states competing with each other'.[11] Two associated themes that dominated Kissinger's 'life work and thinking' – and demonstrated by differences in how world leaders have approached global issues – relate to 'the contest in diplomacy between the realists and the idealists' and 'the principle of non-interference'.[11] While British diplomat Jonathan Powell, who reviewed Kissinger's book *World Order* several years ago, commended it for its 'historical perspective,' he could not agree with Kissinger's premise that 'a new world order requires us to go back to the systems of the 18th and 19th centuries'; Powell concluded that what was needed was 'a new system of our own rather than returning ... to the world order of the past'.[12]

Propositions for sustainability

In light of Powell's comments, there can be little doubt that policy makers engaged with the implementation of the UN SDGs are already climbing a very steep cliff where progress in certain areas of human activity can be undermined in an instant by foolish or careless thinking. The UN-2030 'Transforming our world' agenda commendably calls for the end of 'poverty, hunger, inequality' and protecting the Earth's natural resources by 2030.[13] However, as we are witnessing daily, many parts of the world continue to be in turmoil, as the 'path of escalation and provocation'[14] continues unabated – despite Western sanctions and televised malign images that are as completely incomprehensible as they are deeply reprehensible.[15] With so much at stake and so few answers to contain strategic geopolitical, economic and military imbalances, we – the United Nations? – might be nearing a historical turning-point where a case for a bolder approach to save the world from itself is no longer in the realm of 'wishful thinking' but a matter of necessity – given the state of the planet's biosphere and civilisation.

With the latter in mind, underpinning the 'propositions for sustainability' (see Box 12.1) is the recommendation for a global shift towards a new World Order that would not be distinguished by the continuation of increasingly dangerous West–East, North–South rivalries, threats or power relationships but by 'the kinds of joint decision-making and collaboration needed to solve the world's problems'.[10]

The re-orientation envisioned calls for measures of social progress that impact on people and planet – primarily through education and strategic priorities (see Chapter 10) – and set the stage for improving trust, understanding and relationships with the real potential for 'less conflict within and between societies',[10] and, above all, helping to ensure the sustainability or 'survival' of the planet.[16]

Towards global sustainability

What if those who agreed the Paris Climate Accord late in 2015 (see Chapter 2) and all those they represent adopted a new worldview – that ensures our planet is not only compatible with our needs as human beings but is also compatible with an outer world that meets the needs of our ecosystem? And, *what if* the following propositions were translated into concrete actions at global and national levels based on the values and principles of *inclusion*, *empowerment* and *accountability*?

In his book, *The Age of Uncertainty*, economist and author, John Kenneth Galbraith, probably defined the purpose of 'leadership' best when he said: 'All of the great leaders have had one characteristic in common: it was the willingness to confront unequivocally the major anxiety of their people in their time. This, and not much else, is the essence of leadership.'[17] Another of Galbraith's comments, not easily forgotten – especially at this time when we are facing a

Box 12.1 Ten propositions for global sustainability

What if?

1 *All* 193 NATIONS OF THE WORLD collectively embraced *the vision and aspirations set out in the UN declaration*, 'Transforming our World: The 2030 Agenda for Sustainable Development', in word *and concrete enabling actions?*

2 GLOBAL GUIDELINES were evolved and agreed to ensure *technology* is used *only* for peaceful purposes and in support of the health and well-being of all species and the planet?

3 DECISION-MAKERS treated *migration* as a historical, complex, global human reality that we need to manage collaboratively, compassionately and responsibly with an emphasis on mitigating root causes (socio-political, economic, environmental) while enacting integrated, preventative life-sustaining measures?

4 GOVERNMENT, BUSINESS AND CIVIL SOCIETY collaborated at all levels and adopted a new paradigm – a new world view – to ensure our needs as human beings are compatible with the needs of our outer world – our ecosystem?

5 HUMAN VALUES AND MODERN LIFESTYLES – especially *national vested interests and overconsumption* (energy, water, raw materials ...) – were recognised globally as the leading causes of environmental degradation, eventual loss of resource bases threatening *the sanctity and sustainability of all life?*

6 NATURE was acknowledged as a major source of *human thought processes and feeling* – inspiration and creative learning as well as being a main factor in personal growth and development, thereby contributing to each individual's quality of life, realisation of dreams and aspirations?

7 THE UNIFYING *ONE HEALTH AND WELL-BEING* CONCEPT became the *cornerstone of our education systems and societal institutions*, thereby helping to create a 'more just, sustainable and peaceful world' (UN-2030 Global Goals [SDGs])?

8 HUMANITY'S FUNDAMENTAL ROLE as *frontline custodians of the planet* was recognised and adopted globally, thereby ensuring the true regenerative power of our societies and fostering compassion, trust and goodwill?

9 THE UNITED NATIONS GENERAL ASSEMBLY (UNGA), guided by the human experience over millennia alongside global wisdom, and in a spirit of compassion, kindness, harmony and moral authority, agreed to prioritise and *actively* promote the *values of equality, democracy, tolerance and respect* to bridge divisions between people and bind nations together?

10 *ALL* MEMBERS OF THE UNITED NATIONS SECURITY COUNCIL were held *globally accountable* for their role in maintaining world peace and security – based on a genuine commitment to shared people and planet values while giving a permanent voice to regions with the fastest population growth and social and economic disparities (e.g., Africa, India, Middle East, SE Asia – i.e., close to 6 billion people vs about 2 billion by 2030)?

'constellation of unprecedented challenges to peace and security'[18] – relates to weapons of mass destruction (WMD). Galbraith's assessment should be a wake-up call for all leaders: 'A nuclear war does not defend a country and it does not defend a system ... not even the most accomplished ideologue will be able to tell the difference between the ashes of capitalism and the ashes of communism.'[19] The stockpiling by at least nine nations – Britain, China, France, India, Israel, North Korea, Pakistan, Russia and the US of 'lethal arsenals of nuclear, chemical, and biological weapons and the materials to produce them'[18] continues to present a very real danger to the world.

In light of the potential for nuclear war involving a number of rogue states, it is becoming imperative that approaches other than direct confrontation be found. A University of Cambridge Institute for Sustainability Leadership study reached similar conclusions emphasising, as a top priority, the need for 'the cultivation of a "global" and a "systems" mindset, developing skills of open-mindedness, inclusivity, long-term and systemic thinking, and navigating complexity without trying to artificially reduce it'.[20]

While academic perceptions on leadership and broad-based recommendations are commendable, the question remains: To what extent do they or can they move us forward towards meaningful or constructive change? The answers lie in both addressing immediate dangers while analysing, as Kissinger did, the root causes why it is so difficult to tackle the economic and geopolitical problems that demand solutions that increasingly seem to be in short supply.

In her informative article, 'Globalisation is today's disease. Marxists diagnosed it. We capitalists must cure it',[21] Janet Daley, journalist and Research Fellow at the UK Centre for Policy Studies, tries to do just that. She traces the steps since the late 1980s to today which have been instrumental in the 'free movement of global capital and labour' responsible for transferring power 'from government to companies'.[21] Daley's article is punctuated with words such as 'revanchist nationalism', 'totalitarian capitalism', 'wealth production', 'technological advances' and 'globalisation-freeing of capital (and supposedly, people) from national boundaries and local limitations', thereby allowing the 'interests of corporate capital to supersede the power of governments'.[21]

The disruptive effects on the electorate and on governments have been substantial, Daley says, with 'inchoate rage against "establishment" politicians that are threatening to disrupt the stability and democratic integrity of the West', including Italy, Germany, Sweden, the US – 'all demanding that their own politicians lead them out of this despair' while 'in truth, economics is no longer under the control of governments'.[21] Her plea – 'Somebody, somewhere had better come up with a way out of this, before it becomes irreversible'[21] – might prompt government policy makers and communities at large to reflect on the mindshift advocated in Box 12.1 as well as to 'unpack' ways forward as outlined in the propositions (Box 12.1).

For instance, Professor Rana Mitter, Director of the Oxford China Centre, points out that the success of President Xi's regime will depend on whether the China economy 'endures or falls'.[22] The President's priorities include fighting corruption and 'dealing with local government debt and trying to move from export-oriented growth to domestic consumption' along with promoting 'his personal keynote project, the Belt and Road Initiative, the ambitious plan to bring $8 trillion worth of infrastructure investment to a swathe of territory between western Europe and southeast Asia'.[22] This ambitious project could present an ideal opportunity for China to lead on *One Health* and commit to proposed global mindshift as well as lead on the policies outlined in Box 12.1.

Finding common ground across perceived ideological divisions could be the first step towards a 'more just, sustainable and peaceful world'[13] and has also been advocated by Her Royal Highness Princess Lamia bint Majed Al Saud, Secretary-General of the philanthropic organisation, Alwaleed Philanthropies. Having always believed that 'good communication and education are key to bridging cultural divides' and despite 'horrific events across the world', she firmly believes that Islamic and Western cultures 'must work harder than ever to realise that we have more to share than sets us apart, and an incredible amount to learn from each other'.[23]

Former US president John F Kennedy well recognised the bond that holds the human species together. In his memorable speech at the 10 June 1963 Commencement Address at the American University in Washington DC more than a half century ago (a few months before his assassination) Kennedy pleaded 'not merely peace for Americans but peace for all men and women – not merely peace in our time but peace for all time'.[24] The fundamental importance he placed on safeguarding our children's future should continue to be considered the top priority for all global, national and local leaders in this century.

> So, let us not be blind to our differences – but let us also direct attention to our common interests and to the means by which those differences can be resolved. And if we cannot end now our differences, at least we can help make the world safe for diversity. For, in the final analysis, our most basic common link is that we all inhabit this small planet. We all breathe the same air. We all cherish our children's future. And we are all mortal.[24]

For some it may seem that the world has not made a lot of progress since the 1960s – in view of ongoing conflicts, global warming, poverty and migration. However, Johan Norberg in his book *Progress: Ten Reasons to Look Forward to the Future*,[25,26] echoing Steven Pinker's book *Enlightenment Now*,[3] concludes otherwise. In Norberg's view 'we are witnessing the greatest improvement in living standards ever'.[25] As evidence, he refers to life expectancy which has 'increased more than twice as much as it did in the previous 200,000 years' and poverty: since 1990–1992, 'the number of hungry people has been reduced by 216 million'[25] – although it should be said that about the same number today are overweight or obese.

Perhaps Norberg's most poignant example of how things have improved – at least in some parts of the world? – is a comparison of the life of a 'typical 10-year-old girl born 200 years ago'[25] (say in 1818). She would have been

> stunted and skinny due to chronic undernourishment and recurring famine ... Her surroundings would have been littered with garbage and faeces. She could be taken away by tuberculosis, cholera, smallpox or measles – or starvation ... and not expected to live longer than around 30 years. She would have had five to seven siblings, and she would have seen at least one or two of them die. She would not receive any schooling, and would have been put to work at an early age, perhaps as a domestic servant ...[25]

Moreover, Norberg highlights, 'That same 10-year-old today, even if she lives in one the world's poorest countries, has better nutrition than a girl in the richest countries 200 years ago.'[25]

Norberg is also right that many of the problems we now face have always existed – 'war, crime, disasters and poverty'[25] – although the global population has only grown exponentially since the late 1940s from around 1.5 billion to close to 8 billion today. We may also question his assertion that all problems are 'rapidly declining'.[25] Nevertheless, his observation that our 'amazing progress' has resulted in 'millions of people [being] given the freedom to improve their own lives, and in so doing [they have] improved the world' seems accurate along with his overall deduction that human progress could not have been imposed from the top down by any 'leader or institution or government', while, arguably, being 'humanity's greatest achievement'.[25]

Norberg's overall view on human progress tends towards the ideal as do the UN-2030 SDGs or Global Goals. Treading a circuitous path with a view to achieving the 17 SDG goals and targets, we may need to be mindful that the 'most important advances ... the qualitative leaps' are also perhaps 'the least predictable'[27] and humankind must depend to a large extent on avoiding 'small errors in early steps' as these can 'snowball into large errors in later steps'.[28]

According to Dr Pindai Sithole, Director and social research and evaluation consultant in development at the Centre for Development in Research and Evaluation International Africa (CeDRE Africa), specific elements of leadership capacity play a central role in achieving the SDGs – especially in terms of the extent to which leadership impacts on 'positive change/transformation of people's lives'.[29]

Along with governance[30] and ethical leadership,[31] Sithole asserts that 'human resources management' and 'results-based monitoring and evaluation'[29] need to be mainstreamed in the SDG evaluation framework. In making this argument, he points to the underlying causes for what 'has become to be known as the resource curse theory' – that is, 'where resources which should or [would be] expected to bring blessings to people's lives turn out to cause untold suffering instead and this is no doubt indicative of unethical leadership.'[29]

Sithole's main message – which cuts across all SDGs – is that 'actions of an individual or groups of individuals should always be steered to bring quality of life to people'.[29] The implications of this resolve for safeguarding the health and well-being of the planet – humans, animals, plants, the environment –

supported by lifelong learning, seem equally clear, especially in our efforts to sustain a planet that is at considerable risk.

This view should not be lost on world leaders and reminds us of President Macron's speech to a joint session of the US Congress after meeting with President Trump during a three-day visit. In his remarks President Macron commented that there was no 'Planet B', and touched on four key priorities to ensure a 'more just, sustainable and peaceful world': restricting the proliferation of nuclear weapons; wider adoption of free and fair trade; abandoning protectionism and isolationism; and maintaining the pivotal role of the US in 'fighting for its values overseas'.[32] Perhaps his strongest message to the US Congress was that now is 'the time of determination and courage' as 'what we love is in danger' and that 'closing the door to the world will not stop evolution of the world'.[32]

References

1 Silva R. We can plug Silicon Valley's moral void. *The Times*, 25 February 2018. www.thetimes.co.uk/article/we-can-plug-silicon-valleys-moral-void-6kdvmwqn2 (accessed 3 July 2018).
2 Popper K. *The Open Society and Its Enemies*. Princeton, NJ: Princeton University Press; 2013.
3 Pinker S. *Enlightenment Now: The Case for Reason, Science, Humanism, and Progress*. London: Penguin Books; 2018.
4 Russell J. The march of progress can go into reverse. *The Times*, 15 February 2018. www.thetimes.co.uk/article/the-march-of-progress-can-go-intoreverse-frhq2gr8l (accessed 19 February 2018).
5 Massie A. Democracy may not be dying, but it is sick. *The Times*, 8 March 2018. www.thetimes.co.uk/edition/comment/democracy-may-not-be-dying-but-it-is-sick-zr99m9gxj (accessed 17 July 2018).
6 Worldodometers. *Real time world statistics*. www.worldometers.info/ (accessed 20 October 2017).
7 Moyn S. Hype for the best. *The New Republic*, 19 March 2018. https://newrepublic.com/article/147391/hype-best (accessed 25 March 2018).
8 Rees M. Time has not solved the mystery of creation. *The Daily Telegraph*, 2 January 2018. www.telegraph.co.uk/news/2018/01/02/time-has-not-solved-mystery-creation-sure-humanity-not-end/ (accessed 17 July 2018).
9 United Nations. *World economic and social survey 2013: Sustainable development challenges*. www.un.org/en/development/desa/publications/world-economic-and-social-survey-2013-sustainable-development-challenges.html (accessed 28 October 2017).
10 Rogers DS, Duraiappah AK, Antons DC, Munoz P, Bai X, Fragkias M, *et al.* A vision for human well-being: Transition to social sustainability. *Current Opinion in Environmental Sustainability* 2012; 4: 1–13.
11 Kissinger H. *World Order: Reflections on the Character of Nations and the Course of History*. New York: Penguin; 2015.
12 Powell J. Henry Kissinger's World Order, review. *The Daily Telegraph*, 13 September 2014. www.telegraph.co.uk/culture/books/11092958/Henry-Kissingers-World-Order-review.html (accessed 12 June 2017).
13 United Nations. *Transforming our world: The 2030 agenda for sustainable development*. https://sustainabledevelopment.un.org/post2015/transformingourworld (accessed 3 July 2018).
14 Luhn A. Putin: My nuclear bombs cannot be stopped. *The Daily Telegraph*, 2 March 2018: 12.

15 Agence France-Presse. Watchdog investigates alleged Syria chemical attack. *Daily Nation*, 9 April 2018. www.nation.co.ke/news/world/Watchdog-investigates-Syria-chemical-attack/1068-4379070-k0jg5mz/index.html (accessed 3 July 2018).
16 Likhotal A. New paradigm quest. *Cadmus* 2015; 2(4): 43–47. http://cadmusjournal.org/node/458 (accessed 25 October 2017).
17 Galbraith JK. *The Age of Uncertainty*. 3rd ed. New York: Houghton Mifflin Harcourt; 1979.
18 Reed L. *Weapons of mass destruction*. www.hampshire.edu/pawss/weapons-of-mass-destruction (accessed 20 February 2018).
19 Galbraith JK. *The Ashes of Capitalism and the Ashes of Communism*. New York: Houghton Mifflin Harcourt; 1983.
20 British Council. *Global definitions of leadership and theories of leadership development: Literature review*. www.cisl.cam.ac.uk/publications/sustainability-leadership/global-definitions-of-leadership (accessed 20 February 2017).
21 Daley J. Globalisation is today's disease. Marxists diagnosed it. We capitalists must cure it. *The Sunday Telegraph*, 11 March 2018: 20.
22 Mitter R. Stand back, the red emperor has lift-off. *The Sunday Times*, 4 March 2018: 16–17.
23 Majed Al Saud Lb. *HRH Princess Lamia Bint Majid AlSaud receives the first of its kind 'Change Maker Award' in London*. www.alwaleedphilanthropies.org/hrh-princess-lamia-bint-majid-alsaud-receives-first-its-kind-"change-maker-award"-lo (accessed 16 December 2017).
24 John F Kennedy Library Foundation. *Commencement address at American University, Washington, D.C., June 10, 1963*. www.jfklibrary.org/Research/Research-Aids/JFK-Speeches/American-University_19630610.aspx (accessed 27 January 2018).
25 Norberg J. *Progress: Ten Reasons to Look Forward to the Future*. London: One World; 2016.
26 Norberg J. Things have only got better. *The Sunday Times*, 11 September 2016: 31.
27 Brainy Quote. *Martin Rees quotes*. www.brainyquote.com/quotes/martin_rees_643568 (accessed 20 February 2018).
28 Lee TB. *Will artificial intelligence destroy humanity? Here are 5 reasons not to worry*. www.vox.com/2014/8/22/6043635/5-reasons-we-shouldnt-worry-about-super-intelligent-computers-taking (accessed 29 October 2017).
29 Sithole PM. *Results-based leadership as a vital consideration in the evaluation of SDGs*. http://idev.afdb.org/sites/default/files/documents/files/Results-based%20leadership.pdf (accessed 25 January 2018).
30 Lueddeke GR. Toward holistic governance in an interdependent world [editorial]. *South Eastern European Journal of Public Health* 2017; 9: 1–4. www.seejph.com/index.php/seejph/article/view/176 (accessed 20 December 2017).
31 Laaser U, Schröder-Bäck P, Eliakimu E, Czabanowska K. [The One Health Global Think-Tank for Sustainable Health & Well-being (GHW-2030)]. A code of ethical conduct for the public health profession. *South Eastern European Journal of Public Health* 2017; 9: 1–26. www.seejph.com/index.php/seejph/article/view/177 (accessed 6 January 2018).
32 Riley-Smith B. Emmanuel Macron rebuts Donald Trump on climate change and Iran nuclear deal in Congress address. *The Daily Telegraph*, 25 April 2018. www.telegraph.co.uk/news/2018/04/25/french-president-emmanuel-macron-addresses-us-congress watch/ (accessed 29 April 2018).

13 The promise of Generation Z

> Sometimes it falls upon a generation to be great. You can be that great generation. Let your greatness blossom.
>
> (Nelson Mandela, former president of South Africa)

Johan Norberg's optimism about the present is refreshing but not completely shared by at least one UK group that is feeling the effects of a world in perpetual fast-forward mode, collectively known as Generation Z (Gen Z).[1] This cohort, which follows Generation X, the baby-boomers and Generation Y (also known as the millennials), generally includes those born after 1995 or those just entering adulthood – about 2 billion worldwide with (unlike Europe, North America, Japan and China) a majority of people in African, Middle Eastern, Southeast Asian and many other countries under the age of 30!

Journalist Helen Rumbelow talked to about half a dozen people aged 20 or over about the things that worry them. Highlighting that they are 'a unique experiment in human history' with 'all the problems of their older millennial cousins', she shared their concerns about 'never buying a house or getting a fair crack at social mobility', 'growing up on social media' and feeling 'connected but very, very small'.[2] For some the anger about their life is very real:

> We are living in a world with a lot of crap going on. We see that, and know we are the ones who will be picking up the pieces of what's happening now, with the adults now chucking the problems down to the next generation.[2]

Globalisation itself is perceived as a new challenge, as one respondent lamented: 'From the beginning you are competing with the world, not just with your city or neighbourhood. That's pressure.'[2] Although many 'plan to change the world for the better', there may be an equal number 'who have no idea how they ended up where they are now or how they can get to where they want to be'.[2]

The feelings of uncertainty expressed by some UK members of Gen Z contrast with responses to a survey of their younger American cousins. The study, involving 1,000 participants aged between 13 and 17, was conducted by *sparks*

& honey, a US trend-setting and culturally-focused media company, and asked for views on the future of work, money, technology and education.[3]

Figure 13.1 maps major events that impacted on previous generations' cultures from 1914 to 1997, and includes the present Gen Z timeline 1998–2025. Factors influencing Gen Z are the great recession (2008 financial crisis), the war on terror, the war on climate change and changing lifestyles, especially the belief in making things better.

In introductory comments to the *sparks & honey* report, the authors highlight that Gen Z are used to instability and 'having witnessed their Millennium siblings flail', they are committed to 'doing things differently' and are 'constantly searching for ways to hack life and work'.[3] In addition, they tend to be 'resourceful, creative, humble, always connected' and entrepreneurial; they are are drawn to 'perpetual reinvention' and by 2025 see 'non-linear, overlapping paths of education and work', expecting 'to have multiple careers before they hit 30'.[3] However, caught in a 'breakneck speed of culture', the report concludes that Gen Z may be the last generation 'as we know it': 'Tomorrow will be less about what a difference a generation makes but more about what a difference a day makes'.[3]

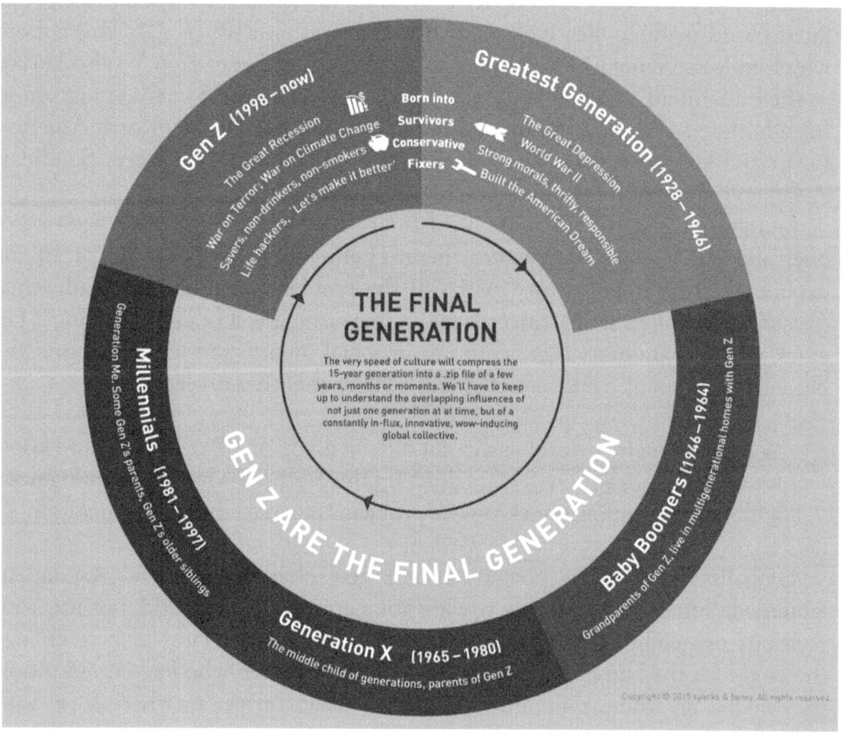

Figure 13.1 Building a vision of Generation Z.

Source: *sparks & honey*. *Generation Z 2025: The final generation* (2015). https://reports.sparksandhoney.com/campaign/generation-z-2025-the-final-generation.

The study also potentially alerts us to the biggest health problem for Gen Z – drawing the line between living in an 'on' and 'off' environment, leading many to seek 'digital detoxification'.[3] In 2015, at the time of the report, it was estimated that 24 million Chinese people were addicted to the internet and about 16 per cent of young adults 'showed signs of on-line addiction'.[3] Psychologist Suzanna Lachmann's observation makes the potential health impact clear: 'It's hard to appreciate where you are when you're perpetually distracted by where you're not.'[3] Beyond diversion and lack of physical vs virtual interaction there may lurk a bigger problem, voiced by a 12-year-old girl: 'If we can control computers with our brains, they can control our memories as well'.[3]

In term of education – with implications for Chapter 10 – Gen Z recognise that going to college is only one option and that they 'want a hand in their own education' (e.g. to design their own courses) and they seek 'learning opportunities everywhere'.[3] For them age is no barrier to education.

Taken as a whole, and based on the US Generation Z study, the most important features that characterise this group include:

- shifting from the 'me to we' re-orientation by 2025
- decomposing 'diversity' as a term 'in favour of simply being human'
- recognising that 'the one thing we have in common is our planet' – that 'caring for the ground beneath our feet is an acute issue'
- putting energy in re-using resources vs creating new ones.[3]

Gen Z and socioeconomic dimensions

Findings of the 'World Happiness Report 2018' (WHR),[4] based on 'life evaluations around the world', point to the need to reflect on the hard realities of life today and where we might be heading in the future. Using the terms 'happiness' and 'subjective well-being' interchangeably, the World Happiness survey, drawing on 'the pooled results from Gallup World Poll surveys from 2015–2017', ranked 156 countries by their happiness levels and 117 countries by the happiness of their immigrants.[4]

Respondents were asked to value their lives today on a scale of zero to ten. All top countries (see Table 13.1) place 'high value on key variables that have been found to support well-being: income, healthy life expectancy, social support, freedom, trust and generosity'.[4]

Demonstrating a distinct North vs South global divide, most countries are below 6.00 (the mean world ranking is 5.26). Regional ratings in descending order are as follows: North America, Australia and New Zealand (ANZ), 6.95; Western Europe, 6.63; Latin America and Caribbean, 6.19; Central and Eastern Europe, 5.84; Commonwealth Independent States, 5.46; East Asia, 5.34; South-east Asia, 5.28; Middle East and North Africa, 5.00; Sub-Saharan Africa, 4.44; and South Asia, 4.35. Forty-nine nations fall below 6.00 and 58 are below 5.00 while 17 lie below 3.00. Thirty-seven countries did not participate in the survey (out of 195 countries in the world today, 193 are member states of the UN and

Table 13.1 The world's happiest and least happy countries, 2018

Happiest countries	*Least happiest countries*
1 Finland: 7.63	147 Malawi: 3.58
2 Norway: 7.59	148 Haiti: 3.58
3 Denmark: 7.55	149 Liberia: 3.49
4 Iceland: 7.49	150 Syria: 3.46
5 Switzerland: 7.48	151 Rwanda: 3.40
6 Netherlands: 7.44	152 Yemen: 3.35
7 Canada: 7.32	153 Tanzania: 3.30
8 New Zealand: 7.32	154 South Sudan: 3.24
9 Sweden: 7.31	155 Central African Republic: 3.08
10 Australia: 7.27	156 Burundi: 2.90

Source: Helliwell J, Layard R, Sachs J. *World Happiness Report 2018*. New York: Sustainable Development Solutions Network; 2018.

2 countries, the Holy See and the State of Palestine, are non-member observer states).

Many past and present reviews of happiness and well-being have been and continue to be based on GDP – 'the total market value of all final goods and services produced in a country in a given year'.[5] While GDP as an average measure is 'a good proxy for well-being' (see Table 13.2), the measure 'ignores the asymmetrical distribution of wealth in a country and continues to correlate wealth and well-being as complementary and generally benign measures'.[6]

As the data in Table 13.2 illustrate, the correlation between GDP and WHR scores and rankings is weak. The US has the highest nominal GDP but is placed 18 in the WHR. Costa Rica is rated 79 economically but has a 7.02 happiness score and a ranking of 13. Only Canada is in the top ten of both GDP and WHR rankings.

These observations are consistent with research that tells us happiness is not always closely associated with income or other objective indicators of well-being such as physical health. In fact, Amartya Sen, acclaimed philosopher and proponent of social justice, has argued that what is most important is providing 'the freedoms and capabilities that allow each person to achieve what will contribute to his or her own wellbeing'.[7] This may place less value on material wealth and shift from economic focus towards 'equality in social relations, social trust in most other people, and degree of democracy; and safety of the area in which one lives'.[6]

It is clear that many of the countries surveyed would fall short of the latter along with not meeting some of the six WHR 2018 criteria for life satisfaction. It may also be apparent that the three major health problems threatening some of the leading economic nations – 'obesity, the opioid crisis, and depression' – are indicative of a bigger problem, the deterioration of emotional health, and should ring alarm bells for all of us – especially in the US, where 'the prevalence of all three problems' is 'growing faster and further than in most other countries'.[4] Among other findings, the WHR highlights 'the importance of social

Table 13.2 Top ten world economies (2018) and correlations between Gross Domestic Product (GDP) and World Happiness Report (WHR) scores/rank

Country	*Rank*	*GDP (US$-trillions)*	*GDP (% of world share)*	*WHR Score (10)*	*Rank (N = 156)*
US	1	20,199,960	23.9	6.88	18
Japan	3	5,063,130	6.00	5.91	54
Germany	4	3,934,810	4.66	6.96	15
France	5	2,765,600	3.28	6.48	23
UK	6	2,661,230	3.15	6.81	19
India	7	2,654,170	3.15	4.19	133
Brazil	8	2,199,720	2.61	6.41	128
Italy	9	2,048,990	2.43	6.00	47
Canada	10	1,763,490	2.09	7.32	10
Costa Rica	79	0.61,064	0.07	7.02	13

Source: Statistics Times. *List of countries by projected GDP*. http://statisticstimes.com/economy/countries-by-projected-gdp.php; Helliwell J, Layard R, Sachs J. *World Happiness Report 2018*. New York: Sustainable Development Solutions Network; 2018.

Note
WHP – World Happiness Report.

factors in the happiness of all populations, whether migrant or not' and emphasises that happiness depends on 'the greater warmth of family and other social relationships', an aspect that could help to explain Costa Rica's high global WHR position (see Table 13.2).[4]

On a global level the question of prosperity – why some nations prosper while others do not – has been investigated by researchers from Germany, Britain and New Zealand who analysed 44 Eurasian countries. Starting with each nation's UN Human Development index score, the study compared income, life expectancy, education against geography, faith, systems of government and a quality called 'deep cultural ancestry' – 'the vertical relationships of descent between cultures'.[8] The study's findings indicate that 'recent cultural history', in particular the doctrines of Marxism–Leninism, are 'more important than deep cultural factors in current human development', contributing, as examples, to low life expectancy (especially in the 1970s and 1980s) that can be attributed to 'high alcohol consumption, smoking and poor workplace safety, as well as low quality diet and living conditions associated with lower income levels'.[8] The researchers conclusion that cultural ancestry is 'not a predictor of human development' could be a positive finding as development outcomes are 'more likely to be amenable to policy change'.[8]

Gen Z and culture change: paying attention to a growing 'global voice'!

Perhaps our best hope in the next few decades lies with culture change and supporting Gen Z – 'the fixers'[3] – by advancing 'global public goods' (GPGs)[9] (e.g. basic human dignity, education/health care, global peace) and helping them 'along the way to live, learn and earn'[10] and their resolve 'to make the world work better'.[3] The importance of this emerging group to the health and well-being of people and their influence on sustaining the planet – transcending political divides! – should be undeniable.[11]

For one thing, Gen Z 'will become the largest consumer group in the US and Europe by 2020' making up '40% of all the population' and, as Ken Hughes, a consumer behaviourist, put it in a blog: 'The millennials might be driving the car, but the road ahead is very much Gen Z'.[12] Along with the characteristics drawn from the Gen Z study in the US, this group has grown up 'knowing instant, knowing immediate, knowing *now*', not in 'a self-centered way' but 'simply hard-wired for everything to work seamlessly'.[12] In respect of the shared economy 'they demand access to the things they need, but they don't seem to really ever feel that inner desire to possess them as much as previous generations'.[12]

In addition, less constrained by convention, Hughes says, 'Gen Z are seeking new freedoms' about 'being themselves, not part *of* a system but the designer of their own', as well as financial freedom – they are likely to have 'several completely different careers'.[12] Their online presence has given them the self-confidence that 'they have a voice, a global one'.[12] Moreover, it is likely that they will become 'the most pluralist generation of all time', having 'curated an impressive digital social network that is most often international, and less skewed by any one colour, race, creed or gender'.[12]

In turn, it is likely that the younger generation in other countries will reciprocate in kind. As examples,[13] in 2018 the average age of the population in Africa is 19.5 (total population – 1.2 billion) with 15 countries where one-half of the total population is under the age of 18. Uganda is a case in point with a population of 41.5 million, a median age of 15.8 years and a literacy rate at 78.4 per cent, recently ranked as 'the world's most entrepreneurial country'.[14] Uganda's development strategy – 'addressing bottlenecks by building hydropower plants, roads, railways, water transport connections and a new airport as well as tackling the lack of electrical power'[14] – could be an exemplar for other developing nations. Its 'reliance on the private sector as the main engine of growth [has] helped to cut the poverty rate by more than half' and 'shifting away from a predominantly subsistence-based agricultural model to an industrialised economy, creating better-paid jobs and boosting productivity'[14] should benefit everyone and bodes well for the future of Gen Z.[3]

Similarly, 58 per cent of Saudi Arabia's population (total – around 34 million) is under the age of 30 while 41 per cent of India's population (total – 1.35 billion) is below the age of 20. With a population of around 656 million (8.59 per cent of the world's population) – more than twice that of the US – the median age in southeast Asia is 28.8 years. In Indonesia (total population – 257 million) around half of the population is below 30 years of age. The population in the Philippines is about 105 million and the median age is 23.5.[13]

In contrast, the median age in China has been steadily rising from around 19.2 in 1979 and projected to rise to 38.7 in 2020, 45.7 in 2035 and about 51.1

in 2100.[15] Along with India, where over 41 per cent of people (around 1.3 billion) are aged below 30, China also has the youngest population in terms of size with the number of people aged below 30 at around 500 million. In both nations the extent to which the Gen Z cohort will be playing a central role in the future remains uncertain but evidence in India suggests that the young are the most aspirational and are leading the call for change.

Reforms are also on the horizon in China. President Xi Jinping's recent 'clear message that China is safeguarding globalisation by announcing specific measures and a timetable in opening the country's door wider to the rest of the world' signals recognition of the need to meet 'people's aspirations for development, innovation and a better life'.[15,16]

With Western sanctions, 'no new growth model for the 21st century', 'rubristic rearmament', exhausted reserve funds, a 'GDP smaller than Texas',[17] coupled with population decline,[13] Russia, according to journalist Ambrose Evans-Pritchard, appears to be facing a period of 'neo-stagnation'.[18] However, in a rapidly changing world, it is equally possible that the younger generation 'could develop their political voice' and bring about 'a pivotal moment for this generation'.[18] In terms of improving life, Russia's biggest challenge may be best summarised by a 20-year-old student: 'What bothers him most is that many are willing to complain but not look for solutions'.[18]

Taken together, the 'Me to We' mindset – connection to a 'One World Reality,' understanding that 'there is a big world out there and they are only a small part' of it, and the belief not only that 'they have a voice' but they 'can use it for change'[12] – could provide the impetus for essential socioeconomic and geopolitical transformations, nationally and globally. One of Gen Z's priorities could be to place much more importance on what society and the world really need for sustainability rather than continuing to prioritise profits for the few, arguably a legacy of Generation X (1965–1980) and Baby Boomers (1946–1964).[3]

Indeed, Gen Z could be the first generation to take on board an important message conveyed by Yuval Harari, historian and author of *Sapiens*[19] and *Homo Deus*.[20] When, in a BBC radio broadcast, Harari was asked about 'the biggest misconception humanity has about itself', he replied:

> Maybe it is that by gaining more power over the world, over the environment, we will be able to make ourselves happier and more satisfied with life [but] looking again from a perspective of thousands of years … it doesn't seem to make people significantly more satisfied than in the Stone Age.[21]

In light of the push for nationalism and protectionism by some countries, it is noteworthy that it was the younger generation who voted, in the main, against this trend in recent elections and referenda[22,23] and who will certainly play a pivotal role in the decades ahead. As Mary Meehan, 'a cultural scientist' writing for *Forbes* has said, 'Gen Z [or Gen 'We'] is now the face of the planet'.[24] She continued:

> They represent boundary-blurring countries and the reality of our shifting global culture. Already they display a great interest in and tolerance of 'other'. They are more naturally equipped to lead and govern in a multi-power, blended, niche world and will expect such from the workplace and government.[24]

Attending a special event on 'Investing in youth to counter terrorism', organised by the Norwegian and Pakistani delegations and the UN Office of Counter-Terrorism, and held at UN headquarters in New York on 12 April 2018, UN Secretary-General António Guterres underscored the importance of 'investing in youth for the fight against terrorism'.[25] He highlighted that

> If we are serious about prevention, and particularly about preventing conflict, we need to be serious about engaging with and investing in young women and men … who are at risk of disempowerment and alienation, and susceptible to extremist narratives and recruitment.[25]

Noting that 'half the world's population (46 percent) is 24-years-old or younger' and that 'Africa and the Middle East have the highest proportions of young people', Guterres called for young people to be brought 'into the conversation, to enable them to express themselves, to listen to them, invest time and resources in them, and empower them to realize their goals'.[25]

The urgency of the UN Secretary-General's plea has relevance for the more economically advanced nations. For example, the final report of the UK Intergenerational Commission concluded that 'the social contract between young and old is "under threat" and that many people no longer believe the country is delivering on its obligations to different populations, impacting disproportionately on the young'.[26] The report sets out a number of recommendations including a substantial increase in 'public funding for social care', 'reformed taxation' weighing heavily upon those above 'state pension age' and 'boosting employment security' for the younger generation and opportunities for training 'to progress'.[26] From a longer-term policy perspective that stretches beyond the current Gen Z cohort, the Commission highlights the importance of engendering 'engagement in the democratic process for each successive cohort by lowering the voting age and automatically registering young adults attending school, college and university to vote'.[26]

Facing a similar dilemma, as discussed in Chapter 5, Japan is responding with a 'robotics revolution that will transform nearly every aspect of society'.[27] The issue is simply that the nation 'won't have enough workers to support its retirees and will enter a demographic death spiral'.[27]

Finding the right balance between technological innovation (dataism)[20] and strengthening the adoption of basic human values (especially compassion, empathy, trust) that respect the 'community of life' may be the most important challenge facing Gen Z and others that follow in this century. If theses cohorts are able to manage this 'techno-human' paradox, they could very well become more successful than any other previous generation in curbing society's huge appetite for exploitation, brutal competition and conflict. They could help shape a world that strives towards well-being, shared prosperity and social stability for all.[6] They could also be best placed to move us forward by taking on board and debating the ten 'propositions for sustainability' summarised in Chapter 12, thereby helping to achieve the UN-2030 SDGs while creating a 'more just, sustainable and peaceful world'.[28]

References

1 Wallop H. Gen Z, Gen Y, baby boomers – a guide to the generations. *Telegraph*, 31 July 2014. www.telegraph.co.uk/news/features/11002767/Gen-Z-Gen-Y-baby-boomers-a-guide-to-the-generations.html (accessed 25 May 2017).

2 Rumbelow H. We know what you're thinking: vain, anxious and selfie-obsessed. *The Times*, 6 November 2017. www.thetimes.co.uk/article/we-know-what-youre-thinking-vain-anxious-and-selfieobsessed-tjrkfckq8 (accessed 10 November 2017).

3 *sparks & honey*. *Generation Z 2025: The final generation*. https://reports.sparksandhoney.com/campaign/generation-z-2025-the-final-generation (accessed 26 September 2017).

4 Helliwell J, Layard R, Sachs J. *World Happiness Report 2018*. New York: Sustainable Development Solutions Network; 2018.

5 Statistics Times. *List of countries by projected GDP*. http://statisticstimes.com/economy/countries-by-projected-gdp.php (accessed 25 March 2018).

6 Rogers DS, Duraiappah AK, Antons DC, Munoz P, Bai X, Fragkias M, *et al*. A vision for human well-being: Transition to social sustainability. *Current Opinion in Environmental Sustainability* 2012; 4: 1–13.

7 Sen A. Well-being, agency and freedom: The Dewy lectures 1984. *Journal of Philosophy* 1985; 82(4): 169–221.

8 Sookias RB, Passmore S, Atkinson QD. *Deep cultural ancestry and human development indicators across nation states*. http://rsos.royalsocietypublishing.org/content/5/4/171411 (accessed 10 April 2018).

9 Global Policy Forum. *Global public goods*. www.globalpolicy.org/social-and-economic-policy/global-public-goods-1-101.html#a2013 (accessed 23 March 2018).

10 Rudgard O. 'Unhappiest generation in a decade' struggles to cope. *The Daily Telegraph*, 5 April 2018: 9.

11 Lueddeke GR. *Global Population Health and Well-Being in the 21st Century: Toward New Paradigms, Policy and Practice*. New York: Springer Publications; 2016.

12 Hughes, K. Forget millennials, Gen Z is the Future: 5 things to think about. *Guided Selling*, 23 February 2018. www.guided-selling.org/gen-z-is-the-future/ (accessed 28 February 2018).

13 Worldodometers. *Current world* population. www.worldometers.info/world population/ (accessed 11 April 2018).

14 The Report Company. Uganda's road to transformation. *The Daily Telegraph*, 17 April 2018: 1.

15 Statista. *Population in China from 2006 to 2016, by age group (in millions)*. www.statista.com/statistics/250753/number-of-people-in-china-by-age-group/ (accessed 11 April 2018).

16 Baijie, A. Tariffs to fall in further reform push. *The Daily Telegraph*, 16 April 2018: 23.

17 Evans-Pritchard A. Russia roars but its economy can barely sustain its superpower status. *The Daily Telegraph*, 12 April 2018: 2.

18 Ayres S. Russian millennials look to their future: 'If we don't do anything now, we will keep living like our parents'. *Los Angeles Times*, 24 February 2018. www.latimes.

com/world/la-fg-russia-profiles-of-youth20180223-htmlstory.html (accessed 28 February 2018).

19 Harari YN. *Sapiens: A Brief History of Humankind.* London: Harvill Secker (Random House); 2014.

20 Harari YN. *Homo Deus: A Brief History of Tomorrow.* London: Harvill Secker (Vintage); 2016.

21 *BBC Radio 4*. How will humans evolve? [interview with Yuval Harari, 24 August 2016]. www.bbc.co.uk/programmes/p045q3yq (accessed 29 September 2016).

22 Center for Information and Research on Civic Learning and Engagement. *An estimated 24 million young people voted in 2016 election.* https://civicyouth.org/an-estimated-24-million-young-people-vote-in-2016-election/ (accessed 20 October 2017).

23 Intergenerational Foundation. *Younger generations: Why we're angry about Brexit.* www.if.org.uk/2016/08/29/younger-generations-why-were-angry-about-brexit/ (accessed 5 September 2016).

24 Meehan M. The next generation: What matters to Gen We. *Forbes*, 11 August 2016. www.forbes.com/sites/marymeehan/2016/08/11/the-next-generation-what-matters-to-gen-we/#5d8f70073500 (accessed 20 August 2016).

25 Pengying. UN chief asks for investment in youth to counter terrorism. *XinhuaNet*, 13 April 2018. www.xinhuanet.com/english/2018-04/13/c_137107202.htm (accessed 18 April 2018).

26 Intergenerational Commission. *A new generational contract: The final report of the Intergenerational Commission.* www.resolutionfoundation.org/app/uploads/2018/05/A-New-Generational-Contract-Full-PDF.pdf (accessed 11 May 2018).

27 Eberspacher S. Everything you need to know about Japan's population crisis. *The Week*, 11 January 2014. http://theweek.com/article/index/254923/everything-you-need-to-209 (accessed 17 July 2018).

28 United Nations. *Transforming our world: the 2030 agenda for sustainable development.* https://sustainabledevelopment.un.org/post2015/transformingourworld (accessed 20 September 2017).

Epilogue

Homo Deus or Homo Humilem? The choice is ours

In *Sapiens: A Brief History of Humankind*, author Yuval Harari conjectures that 'after the 4-billion-year-old regime of natural selection' – *Homo sapiens* in the twenty-first century are now able to break 'free of their biologically determined limits' in which 'life may be ruled by intelligent design'.[1]

Elaborating this theme in *Homo Deus: A Brief History of Tomorrow*[2] Harari contends – possibly in line with the late Stephen Hawkins' prognostications[3] – that during the course of this century technology (non-conscious algorithms) might seek to control humanism (feelings), assisted by 'genetic engineering, nanotechnology and brain–computer interfaces', possibly severing 'the humanist umbilical cord altogether'.[4] *Homo Deus* might begin to worship a new ideology, 'a bolder techno-religion, Dataism', which 'venerates neither gods nor man – it worships data'.[2] This chilling scenario would require replacing the meaning of human experiences 'inside us' to 'infusing the universe with meaning' because 'dataists believe that experiences are valueless if they are not shared'.[2]

The battle – humanism vs dataism – which already appears to be in full swing, judging by societal, military and Big Tech developments[4] will essentially be between 'biologists and the intelligent design movement'.[2] Initially, Harari observes, dataism will 'probably accelerate the humanist pursuit of health, happiness and power – by promising to fulfil these humanist aspirations'.[2] However, in due course, he speculates, achieving 'immortality, bliss and divine powers of creation' will require processing 'immense amounts of data, far beyond the capacity of the human brain', leading to the inevitable conclusion that 'the algorithms will do it for us' and thereby make 'the humanist projects' irrelevant.[2]

Most will find the latter dystopian outcome for humanity philosophically and morally indefensible.[5] However, while the SDGs are leading us in the right direction, they may fall short, not because of a lack of verbal commitment to the goals by the 193 UN member states but because of our restricted worldview – especially failure to recognise the potential devastating impacts, to people and the planet, of creating global divisions, justifying intolerances and valuing virtual over natural realities, including placing data and the sciences over harmonious human interactions, creativity and enjoyment of life.

There is no doubt that in the long run education (see Chapter 10) remains our best defence against intolerance, armed conflicts, corruption, extremism, digital hegemony and misguided decision making. However, we should be cautious. Education that treats the arts 'as inferior' and pushes students towards pursuing mostly STEM (science, technology, engineering and mathematics) subjects may do society more harm than good in the long run, according to Alice Thomson, a columnist writing from a British perspective.[6] For Thomson, the 'government message', which is not dissimilar to those in many other countries, is that 'the humanities are second rate and "of less economic benefit"'.[6] As a result, subjects such as 'music, art, design, media and drama' are dropped, forgetting that in the future – as is the case already – 'many jobs that rely on rote learning, formulae, and basic questions will be done by artificial intelligence'.[6] The obsession with 'focusing on maths and science', she notes, also ignores the fact that in the UK, 'The creative industries are worth £92 billion … more than the oil, gas, life sciences, automotive and aeronautical industries combined' and that 'What makes humans different is the ability to perform intuitive leaps, to collaborate and to create' through 'music, art, design, media and drama'.[6]

It is unlikely that STEM subjects with their emphasis on 'hard' disciplines[7] and competitiveness will help us create 'a more just, sustainable and peaceful world',[8] place collaboration and diplomacy over alienation or guide our way towards a future that works for all, the overarching UN-2030 SDG aspiration. Modern education must, therefore, ensure not only that the sciences, the arts, the humanities and the physical/experiential are fully integrated in a holistic curriculum but also that the student learning experience is exploratory, multidisciplinary, interdisciplinary and global. This blending of knowledge, skills and creativity is becoming increasingly important. As Thomson observes, 'In the future, if you want an interesting job, you will need to become as unlike a machine as possible, proving you can be innovative, emotionally intuitive, flexible, and collegiate.'[6]

In a commentary on Alice Thomson's article, Sir Simon Rattle, music director of the London Symphony Orchestra gives perhaps one of the most cogent arguments as to why the arts matter – while recognising Thomson's piece as 'a call to action'. He argues:

> … it is essential to the wellbeing and success of our young people that their education prepares them for a future where imagination and creativity are the most important attributes. The future may be uncertain and difficult for the next generation, but unless they have access to a vital cultural education they will be utterly unprepared for what this new world may require. Our children need to have the artistic vitamins that will help to build a better society.[7]

Rattle's insights echo many of the themes in the groundbreaking Development Alternatives report, 'To choose our future',[9] commissioned by the Club of Rome, which, together with 'Five principles of peaceful coexistence' (see Box E), offers alternative development strategies for India to build a more prosperous, vibrant and sustainable future.

> We need to refashion our institutional systems and transform our current attitudes to virtually all aspects of society and the economy – consumption patterns and wellbeing, technology and production systems, enterprise and distributive justice – all of which in the light of today's circumstances and knowledge need to be reoriented to conform to the principles of an inclusive and circular economy. This implies that the poorest and marginalised are put at the centre of economic and social attention and the restoration and regeneration of natural systems become the boundary conditions that must not be transgressed, not just for future generations but also for those of today.[9]

While intended for a society undergoing rapid socioeconomic transformation, the priorities principles and practices espoused could easily apply to many other developing and developed nations today facing essentially identical issues and grappling with the need to find solutions that last.

Box E Development alternatives

'Five principles of peaceful coexistence'
(Panchsheel Treaty – Sanskrit)

Principle 1: Commitment to universality and inclusion
Principle 2: Adopting lifestyles of sufficiency
Principle 3: Designing systemically for efficiency
Principle 4: Building diversity for resilience
Principle 5: Regenerating nature and protecting life.

Khosla A. Development Alternatives. *To choose our future*

In his seminal book, *Collapse: How Societies Choose to Fail or Survive*,[10] 'an epic journey around the globe, through the history of humanity and on to the future, to discover how – when tomorrow comes – we can be survivors', University of California, Los Angeles professor Jared Diamond maintains that, 'the collapses of past societies have modern parallels'[10] and offer us many lessons. Diamond – 'ecologist, geographer, biologist, anthropologist' – reasons that many of today's problems stem from 'countries that are environmentally stressed, overpopulated, or both' and that these are 'at risk of getting politically stressed, and of their governments collapsing'.[10] Being 'desperate, undernourished, and without hope', he contends, people 'blame their governments, which they see as responsible for or unable to solve their problems'.[10] Consequences include conditions that create 'not only chronic internal conflict, but also emigration of political and economic refugees, and wars between countries arising when authoritarian regimes attack neighboring nations in order to divert popular attention from internal stresses'.[10] In the extreme, Diamond contends, 'feeling they have nothing to lose', some turn to terrorism.[10]

What distinguishes us from the 'ancient Maya, Anasazi, and Easter Island' societies, Diamond observes, is that today we are all interconnected and interdependent: what happens in any place on the globe (e.g. social, political, economic, environmental) can have profound repercussions everywhere. For Diamond, the notion that nations can advance their 'own self-interests at the expense of others' or that 'the elite can remain unaffected by the problems of society around them'[10] seems both outdated and morally indefensible. He concludes that societies have two choices that may determine whether they 'fail or survive': 'long-term planning and willingness to reconsider core values'.[10]

Long-term thinking is certainly in order when we consider the state of the planet highlighted in the Introduction to this book and reinforced by the 'ten propositions for global sustainability' outlined in Chapter 12. Its urgency is also magnified when we consider recent statistics pertaining to human population which underscore that:

- Over 800 million people (around 10 per cent of a world population of nearly 8 billion) still live in extreme poverty.[11]
- Five billion people cannot access safe surgical care, with millions dying 'from treatable conditions and injuries, such as appendicitis, obstructed labor, and hernias'.[12]
- About 2.1 billion people lack access to safe, readily available water at home, and 60 per cent, or 4.5 billion, 'lack safely managed sanitation'.[13]
- About 1.1 billion people worldwide now have a mental health problem (especially anxiety and depression) or substance use disorder.[14]
- There is a 'median estimate for future growth that sees the world population reaching 8.6 billion in 2030, 9.8 billion in 2050 and 11.2 billion in 2100'.[15]

Diamond's 'other crucial choice illuminated by the past involves the courage to make painful decisions about values'.[10] That is, deciding 'Which of the values that formerly served a society well can continue to be maintained under new changed circumstances?'.[10] It is in this respect that this book attempts to convey seven fundamental 'take away' messages that place

1. the health and well-being of the planet and the sanctity of life (humans, animals, plants, environment) *over* vested self-interests;
2. compassion, empathy and life-enhancing opportunities for those marginalised in society *over* intolerance, indifference or neglect;
3. human rights and humanitarian values *over* expediency and technology;
4. funding for peace and well-being *over* war and destruction;
5. prevention *over* intervention (e.g. individual health and well-being, conflicts, environment);
6. personal commitment to make the world sustainable *over* the status quo, inertia or complacency; and
7. learning, inspired by imagination and creativity – that cultivates understanding of the world, respect for all life on the planet and hope for a 'more just, sustainable and peaceful future' – *over* ignorance, oppression and the denial of truth.

Dr Alexander Likhotal, former president of Green Cross International and member of the Futures Council, calls for 'true political leadership, prophetic vision and courage' arguing that: 'Change is no longer a mere theory, and it is no longer just an option: it is a reality, a "condition sine qua non" of our *survival*'.[16]

Looking ahead to the remaining decades of this century, we will likely also need to brace ourselves for a world that transcends today's realities: to make a choice between *Homo Deus* or *Homo Humilem*, between 'highly intelligent algorithms'[2] (superhuman or artificial?) and a society that truly cares for the 'sanctity of life' – humans, animals, plants, environment – and continues to make life on the planet possible. Our decisions with the former may determine the outcomes of the latter.

Despite considerable pressure from technology giants, the military and politicians, the deciding factor could simply come down to our collective ethical resolve not only to stay human but also, as Klaus Schwab, founder and executive chair of the World Economic Forum (WEF) concluded, 'to become better humans'.[17] In the 2016 WEF panel discussion on shaping technologies, Schwab highlighted ways forward: 'If we all create self-awareness and, in our own personal and collective lives, work towards improving the state of the world, we would live out our human dimension.'[17]

Following Schwab's global exhortation, at long last we might finally realise two fundamental truths. First, as the late physicist, Stephen Hawking, cosmology's 'brightest star', bluntly put it: 'We are all different … but we all share the human spirit'.[18] And, second, we might begin to understand the fragility of the planet and that unless crucial societal transformations occur – including the prevention of 'nuclear war, global warming and genetically engineered viruses',[19] as well as putting all people and the planet first – the shelf life of *Homo sapiens* could be extremely short.[20]

Returning to the Preface of this book, our best hope for averting the possibility of a dehumanising or dystopic world in the twenty-first century – or worse – may lie with recalling the eloquent words of civil rights leader, Martin Luther King Jr (1929–1968): 'Our lives begin to end the day we become silent about things that matter'.[20,21]

References

1 Harari YN. *Sapiens: A Brief History of Humankind*. London: Harvill Secker (Random House); 2014.

2 Harari YN. *Homo Deus: A Brief History of Tomorrow*. London: Harvill Secker (Vintage); 2016.

3 Seidel J. Artificial intelligence is a risk to humanity says astrophysicist Stephen Hawking. *News Corp Australia Network*, 3 December 2014. www.news.com.au/technology/science/evolution/artificial-intelligence-is-a-risk-to-humanity-says-astrophysicist-stephen-hawking/news-story/a4e3cb173ba00a00afa8e4f893c4dff9 (accessed 29 December 2014).

4 Blakely R. Google workers declare war on AI drone project. *The Times*, 6 April 2018: 30.

5 The Vatican. *Encyclical letter Laudato Si' of the Holy Father Francis on care for our common home.* http://w2.vatican.va/content/francesco/en/encyclicals/documents/papa-francesco_20150524_enciclica-laudato-si.html (accessed 17 July 2018).
6 Thomson A. Children are being turned into mini robots. *The Times*, 16 May 2018: 25.
7 Rattle S. The value of arts education to our prosperity. *The Times*, 18 May 2018: 28.
8 United Nations. *Transforming our world: the 2030 agenda for sustainable development.* https://sustainabledevelopment.un.org/post2015/transformingourworld (accessed 20 September 2017).
9 Khosla A. Development Alternatives. *To choose our future.* www.perspectives.devalt.org/?p=66722 (accessed 17 October 2016).
10 Diamond J. *Collapse: How Societies Choose to Fall or Survive.* London: Penguin Books; 2006.
11 World Health Organisation. World Bank and WHO: Half the world lacks access to essential health services, 100 million still pushed into extreme poverty because of health expenses [news release, 2 June 2018]. www.who.int/news-room/headlines/13-12-2017-world-bank-and-who-half-the-world-lacks-access-to-essential-health-services-100-million-still-pushed-into-extreme-poverty-because-of-health-expenses (accessed 2 June 2018).
12 Advisory Board. The Lancet: Five billion people lack access to proper surgical care [news brief, 29 April 2015]. www.advisory.com/daily-briefing/2015/04/29/five-billion-people-lack-access-to-proper-surgical-care (accessed 29 November 2017).
13 World Health Organisation. 2.1 billion people lack safe drinking water at home, more than twice as many lack safe sanitation [news release, 2 June 2018]. www.who.int/news-room/headlines/12-07-2017-2-1-billion-people-lack-safe-drinking-water-at-home-more-than-twice-as-many-lack-safe-sanitation (accessed 3 June 2018).
14 Roser M, Ritchie H. *Mental health.* https://ourworldindata.org/mental-health (accessed 10 May 2018).
15 UN Department of Economic and Social Affairs. *World population prospects: The 2017 revision.* www.un.org/development/desa/publications/world-population-prospects-the-2017-revision.html (accessed 28 June 2017).
16 Likhotal A. New paradigm quest. *Cadmus* 2015; 2(4): 43–47. http://cadmusjournal.org/node/458 (accessed 26 November 2017).
17 World Economic Forum. *A key challenge of the fourth industrial revolution: Staying human – and humane.* www.weforum.org/press/2016/01/a-key-challenge-of-the-fourth-industrial-revolution-staying-human-and-humane/ (accessed 29 March 2017).
18 Sherrington T. *Hawking: 'We are all different … but we all share the human spirit'.* https://teacherhead.com/2012/08/31/hawking-we-are-all-different-but-we-all-share-the-human-spirit/ (accessed 1 April 2018).
19 Knapton S. Prof Stephen Hawking: Disaster on planet earth is a near certainty. *The Daily Telegraph*, 19 January 2016. www.telegraph.co.uk/news/science/science-news/12107623/Prof-Stephen-Hawking-disaster-on-planet-Earth-is-a-near-certainty.html (accessed 20 January 2016).
20 Federally Employed Women. *Presidential comments.* www.few.org/still-dream/ (accessed 22 June 2018).
21 World's Largest Lesson. *End extreme poverty. Fight inequality and injustice. Tackle climate change.* http://worldslargestlesson.globalgoals.org/ (accessed 17 February 2018).

About the author

George Lueddeke MEd PhD is an educational advisor in higher and medical education and chairs the global One Health Education Task Force for the One Health Commission and the One Health Initiative (Chapter 11). Originally from Canada he has worked in both northern and southern communities as a teacher, researcher, educational manager and developer as well as programme consultant across a wide range of disciplines/professions. In the UK he has held posts in higher education and as senior lecturer in medical education at Southampton University's Faculty of Medicine. As an advisor he has collaborated inter alia with the NHS Kent, Surrey and Sussex (KSS) Postgraduate Deanery in London, the UK Centre for Workforce Intelligence, the UK Care Quality Commission, the Association of Schools of Public Health in the European Region (ASPHER), the Commonwealth Secretariat and the InterAction Council. Most recently, he has contributed to initiatives such as Organisation for Economic Cooperation and Development (OECD) Education-2030, the SEEJPH Public Health Curriculum Project, the ASPHER/WHO/Imperial College professionalisation of the public health workforce, and the South Africa Medical Association (SAMA)–South Africa Veterinary Association (SAVA), among others, on the implementation of *One Health and Well-Being*. In addition, he chairs the One Health Global Think-Tank for Sustainable Health & Well-being' – 2030 (GHW–2030) to encourage multi-sectoral and multi-disciplinary research underpinning the sustainability of planet and people.

Dr Lueddeke has published widely on educational transformation, innovation and leadership. Previous books have included *Transforming Medical Education for the 21st Century: Megatrends, Priorities and Change* and *Global Population Health and Well-Being for the 21st Century: Toward New Paradigms, Policy and Practice.*

Invited as a plenary speaker to different corners of the world, he has presented sessions for organisations such as the UK General Medical Council, the American Medical Association, University 21 (U21) – Health Sciences Group (Millennium Goals – Dublin), the Public Health Association of South Africa, University of Sri Lanka Jayewardenepura Medical School, UK Faculty of Public Health, The Catharina Pijls Lecture on 'Global and European Health' (annual Maastricht Symposium), the Global Health Council, Cambridge University, the UK Veterinary Public Health Association, the World Veterinary Association-World Medical Association (Kitakyushi, Japan), the International Schools Association annual 'Youth Leadership Encounter' conference on 'Worldwide Migration' (UK) and the Arab Health Congress (Dubai).

Want to change the world?
Start by being brave enough to care.

(Cleo Wade, artist and poet. TED Talk, *Women 2017*. To watch the full talk, visit https://archive.org/details/CleoWade_2017W)

A journey of a thousand miles begins with a single step.

Lao Tzu (ancient Chinese philosopher, sixth century BCE)

Index

Page numbers in *italic* denote figures and in **bold** indicate tables.

For Product Safety Concerns and Information please contact our
EU representative GPSR@taylorandfrancis.com Taylor & Francis
Verlag GmbH, Kaufingerstraße 24, 80331 München, Germany